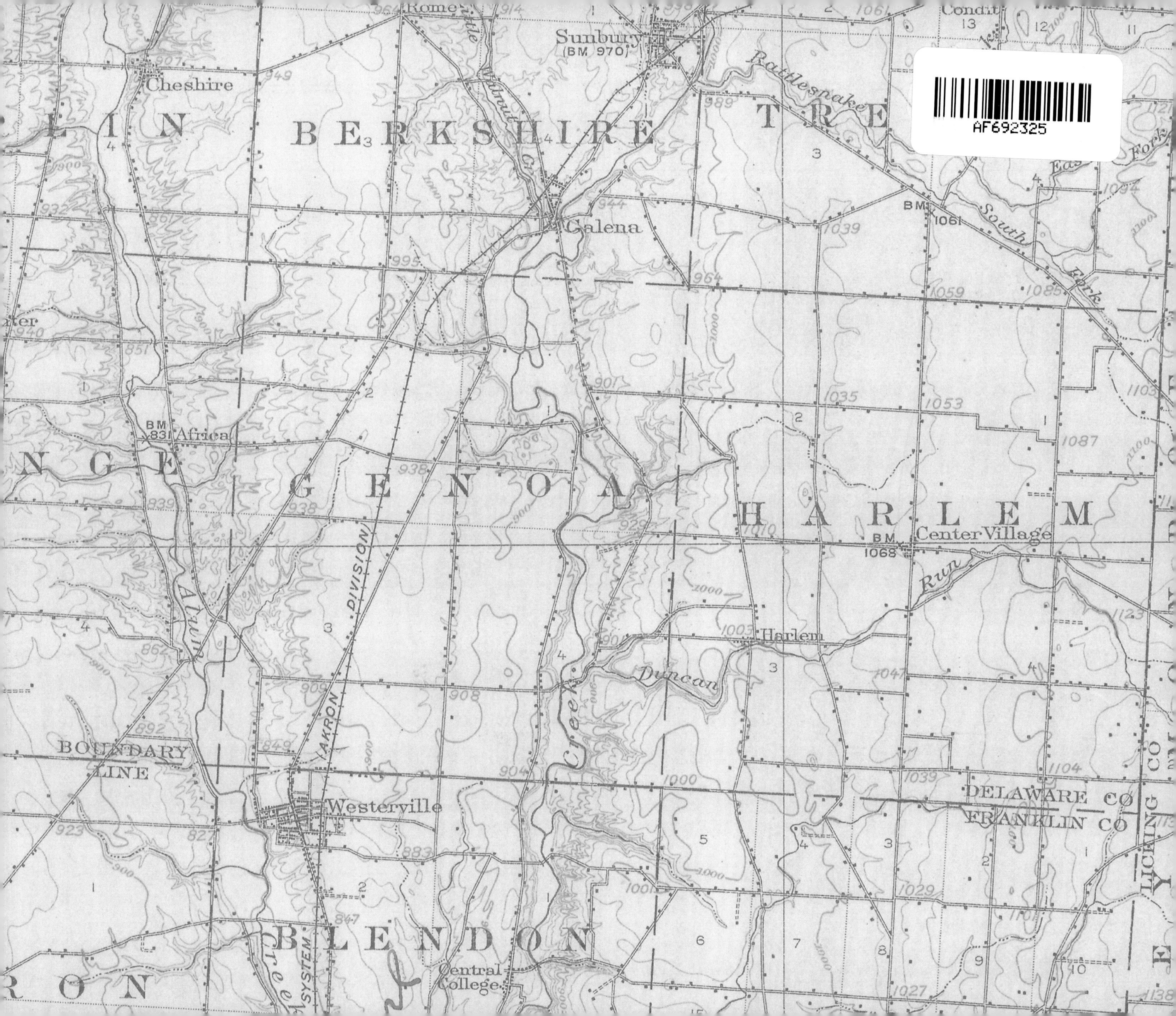
Rome
Sunbury
(BM 970)
Cheshire
BERKSHIRE
Galena
Rattlesnake
South Fork
Africa
GENOA
HARLEM
Center Village
Harlem
Duncan
AKRON DIVISION
BOUNDARY LINE
Westerville
DELAWARE CO
FRANKLIN CO
LICKING CO
BLENDON
Central College

ELENA DORFMAN

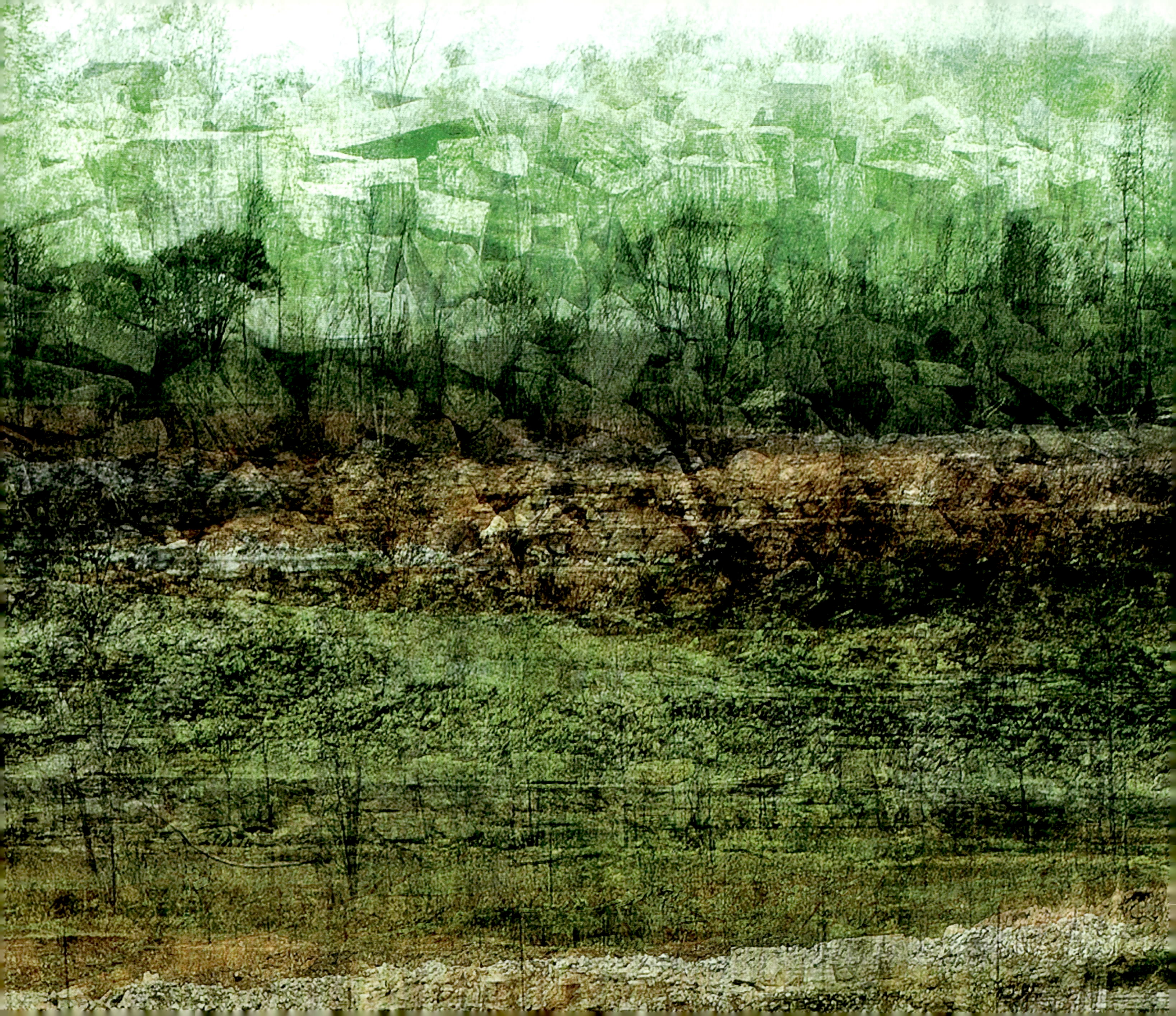

ELENA DORFMAN
EMPIRE FALLING

crump
DAMIANI

in association with

MODERNISM
685 MARKET STREET · SAN FRANCISCO, CA 94105
TUES-SAT 10-5:30 (415) 541-0461 FAX 541-0425
WWW.MODERNISMINC.COM

X 10

Times Ten: Elena Dorfman's Post-Industrial Abstractions

Kevin Moore

Deaths by drowning in quarries are commonplace because quarries are not built for swimming, they are "repurposed" for swimming by heedless adventurers. Quarries are places dangerous and alluring, both natural and man-made, where the ruins of modern industry succumb to organic predation—flood waters, tenacious vines, brave saplings, invading a defunct terrain of collapsed buildings, corroded equipment, and overhead wires. It is thus not surprising that quarries attract an outlaw audience of rebel teenagers, strategic geologists, and intuitive artists. Quarries hold the promise of answers to half-formed questions—historical, scientific, psychological questions—revealed, potentially, in a gashed, abandoned landscape.

A teenager drowned in a flooded quarry in Verplanck, New York, in 2009, and the *Times* coverage of the tragedy, ostensibly warning off trespassers, read more like a travel brochure for extreme tourism, describing "50-foot cliffs and remarkably clear turquoise water 300 feet deep just down from the Indian Point nuclear plant."[1] Quarries such as this one are, of course, sites of obsolescence, little more than legal liabilities to their owners, who see no use in them—only potential lawsuits—and so vaguely fence them off. Yet such sites broaden, as if consciously, toward an industrial afterlife, the promise of new social functions quickening alongside nature's irrepressible display. Beyond the unauthorized uses, quarries are being turned regularly into amusement parks, housing developments, swimming pools, golf courses, tennis courts, such projects officially sanctioned, their raw, not-to-code character—precisely what attracted the teenagers, geologists, and artists in the first place—now paved, landscaped, handrailed.[2] As defunct industrial ruins become exotic, the exotic becomes an attraction in which the dangers are safely cordoned off.

The paving, landscaping, and handrailing might also be seen to be part of a grand effacement—of the past, generally, but of a now seemingly shameful industrial past, specifically. Our current environmentalist sensibility, countenancing with horror the "scars of industry" qua quarries and other ravaged natural sites, seeks to normalize, familiarize, domesticate. A quarry turned into a shopping mall becomes a cleansing of conscience, a recovered "industrial wasteland," though one wonders at such dubious transformations. In our rush to sites of sublime beauty (quarries, arguably, are as sublimely beautiful as they are abject) we tend to spoil them, reducing torrential rivers to waterslides, mountains to Mountain Dew.

It may be helpful to remember that quarrying was once a noble industry, producing the rock that built the nation, its capitals, monuments, and mansions. The quarries in Kentucky and Southern Ohio, in particular, provided the limestone and sandstone for much of the country's most impressive architecture, elaborating a symbolic program of power and permanence: buildings made from solid rock, an ancient geologic history now constituting an emboldened American institutional history—a declaration of permanence, or an illusion thereof.[3] On a much more discreet level, quarries provided the raw material for roadways and fertilizers—foundational bedrock to architecture's ornament. Both uses demonstrate a conquering of nature, the transformation of subterranean chaos into terrestrial order, of raw material into social structure. A hole in the ground points, obliquely, to the state capital, the small town bank, the Bunker Hill Monument, the White House, if one is sensitive to the sprawling cavity at hand.

Elena Dorfman's photographs treat this complex, creeping history by documenting its aspect and complementing its forms. Layers of history may be seen—may be *unearthed*—in *Empire Falling #5* (plate 22). Part of the mystery here is that the swimming pool is drained, revealing a tracery of cryptic symbols: black lines, crosses, and graffiti, this last element migrating, like some exotic mold, up the quarry wall, which serves as a decomposing backdrop to the pool's proscenium. A more literal photographer might simply have documented this evident environmental contrast, deftly framing nature and artifice, or perhaps emphasizing the commonality: architectural ruins. But Dorfman's photograph capitalizes on the complexity of the setting, sparking a radiating analysis, blending nature and artifice, layering one ruin over another, searching the nooks and crannies for evidence of past actions, uses, habitation. Her photograph engages the very process it sets out to record: it is a laborious excavation of events—both natural and human—over time.

Photography and quarries both manifest time. Photography is often said to capture time, but it would be more accurate to say that photography creates awareness of time, setting in motion a meditation on the time that has passed since the making of the photograph and the present moment, and all that has changed or disappeared in between. Photographs, upon their instant of manufacture, become historical documents that attest to a moment in time, ever receding.

Quarries expose time but take a broader view, revealing the strata of millennia, layer upon layer. "Geology is the book of earth time," a noted geologist has said. "The past begins with the oldest, at the bottom, and works its way up to the surface, to now."[4] If a photograph is a single layer of time—if it is the flat surface—then the exposed wall of a quarry is a chronology, a timetable, an account book of time itself, descending below. Yet little is revealed in the way of lived events. A photograph is a flush of information about a given subject at an instant in time; the wall of a quarry is a clock set in nature, keeping track of time, marking its progress. In short, a photograph is a tableau: a picture. The wall of a quarry is a tablet: a compressed solid substance. Between the two, tableau and tablet, time can be seen horizontally or vertically, as a synchrony or a diachrony, an elaborated single moment or a vertebrae of consecutive moments—or, potentially, in a photograph of a quarry, all such time at once.

Photography and quarries—not the strata of rock revealed, but the industrial enterprise that has revealed it—also speak commonly of a specific historical subject, a period in time known as the modern industrial age. Both evidence the modern industrial age as progeny of that age, photography and quarries being machine-driven practices devised to abet progress, i.e., the growth of a national economy and its visual dissemination—one industrial practice attesting to another. Today we may doubt photography's veracity, the medium's so-called objectivity (photographs, we now acknowledge, always have been manipulated),[5] but the modern period's belief in photography as a rational process was part and parcel with a belief in industry as a rational process, its heroic, ameliorating purpose rarely questioned. Early twentieth-century photographs by Lewis Hine, Charles Sheeler, and Margaret Bourke-White captured in silver gelatin the monumental mechanisms of the machine age: steel mills, automobile plants, bridges, skyscrapers. In each case, the material of the photograph matches the material of its subject: cold metal and hard edges, a new kind of beauty based on geometric forms and their rational functions.

With the twilight of the modern industrial age, we see photography and quarries (and other industrial subjects) becoming natural sites of elegy, nostalgia-laden testimonials to a lost heroic past, and this sentiment is certainly, fractionally present in Dorfman's pictures. There is a cemetery-like hush in her images, a sense of abandonment, of depopulation, where the air almost becomes visible as an animus. Other artists have treated similar subjects: Bernd and Hilla Becher's typologies of industrial sites; Vera Lutter's haunting, camera obscura-based images of architecture; Zoe Leonard's dye-transfer typologies of declining storefronts; Simon Starling's platinum prints of mines. In each case, a correspondence is registered between the materiality of the print, the mechanics of the process, and the subject at hand: industrial means depicting (seemingly lamenting) a fading industrial era. Photography stages its own perceived demise.

But this is not Dorfman's purpose. Although industrial motifs appear in her photographs—cables, conveyor belts, sand silos—Dorfman's pictures are distinguished in a crucial, technological way: they are digitally captured and heavily manipulated. Thus their materiality and process enjoin a progressive notion of photographic technology (photographic technologies have always been progressing, it should be emphasized), paralleling current economic processes, seen to be evolving, synthesizing, abstracting. If film technology and the non-manipulated image are now associated with the industrial age, digital technology and a jettisoning of naturalistic vision have come to be associated with the post-industrial age. One form of realism is supplanting another: a paradigm of transparency, revelation, is now being occluded by mediation, fabrication. Photographic film, a physical material used to document specific sites of industry, is replaced by computer bytes, ephemeral units storing visual information for virtual circulation and recombination. Just as the American economy increasingly abandons the specificity and physicality of factories to embrace abstract money schemes and ubiquitous internet-based services, photography is leaving its concrete origins in film and prints to circulate on the World Wide Web.

Thus the abstraction in Dorfman's photographs may be seen to mirror—to represent *realistically*—economic conditions historically endemic in, and symbolized by, quarries. In their day, quarries provided a raw material—stone—which contributed to the foundation of the national economy, generating capital for more elaborate manufacturing industries, such as construction, engineering, architecture, and all the businesses the structures created by those enterprises went on to house. Today's economy is far more abstract. Capital has become finance capital, which is capital built on capital, money generated by money, "liberated" from any specific material or profit-generating industry—an abstraction based on an abstraction.[6] So when Dorfman photographs a quarry and goes on to montage, layer, and recombine images, she is not only capturing in the original images the original conditions of a genesis of capital; she is also representing, in her layering and abstracting of images, subsequent conditions, i.e., the development and circulation of finance capital. And she is creating single synthetic images that are true to her subject: layers of images mimicking layers of stone. "A seam looks like a seam," the artist has noted, referring to the evident montaging of her images. Sometimes the combining is quite evident, as in an image such as *Empire Falling #1* (plate 54), where the layers of information create a shattering, destabilizing effect. Or in *Empire Falling #20* (plate 73), where an inset frame creates a pictorial jolt, grafting tableau over tablet. In other instances, the seams are more seamless: *Empire Falling #21* (plate 78) comprises information from hundreds of digital captures, combined in one large picture to create a density of graffiti—horizontal layers over vertical layers, tableaux over tablet— that is non-existent in real life.

Photography is a fundamentally paradoxical medium. Almost any statement about photography can be turned on its head: Photography depicts in a realist fashion, transcribing the visible world directly; or, photography is fundamentally abstracting in its transcription of the three-dimensional onto a two-dimensional support. Photography shows reality, "truth"; or, photography conjures fantasy, "tells lies." Photography is objective, apolitical; or, photography effortlessly supports any political or commercial use. Photography is a hesitantly realist medium in the sense that the naturalistic surface of a photograph often masks complex social, economic, and political realities, just as public relations firms mask the machinations of corporations, politicians, and celebrities. Reality most often lies beneath the surface, in the tablet.

Paradoxically, a digitally constructed image can deliver more specific information, more photographic detail, than a "straight" or non-manipulated photograph, because it is not held to a single location or instant in time. Rather than being unreal, the digitally constructed image is hyper-real, eructing a cascade of information from many different points of view. Surprisingly, this replicates natural human vision more closely than the single-perspective image, as it allows the viewer to "look around" within the image, much as one looks around—as opposed to looking *into* the rectangle defining a traditional photograph—in everyday life. (Paul Cézanne's paintings were praised for this sort of optical naturalism even though they obviously flouted a Western painting tradition based on one-point perspective.) A digitally constructed image may not be what we're used to seeing, but the way we experience the image is closer to how we see.

Thus photography's claims to realism should be understood as having more to do with the materiality of the photographic process than the ostensible subject depicted. Gustave Courbet's realism was based on heightening an awareness of the real substance of painting—paint, seemingly transformed into stone, in his depiction of the *Stone Breakers* (1850), which also happened to embody a social realism: hard labor. Here technique and subject are seamlessly combined.

Photography in Elena Dorfman's hands enacts the same dialogue, such that what we experience in relation to her pictures is not so much depictive as it is *indicative* of a set of relationships, complexly diagrammed: layers of time; materiality of the photographic process related to the materiality of industry; obsolescence and endurance; the labor in photography related to the labor in the quarry; time expressed simultaneously in instants and in chronologies; economies based on raw substances and on the capital generated from the exploitation of raw substances; the simultaneous depiction of the visible and the invisible; the physical and the immaterial; the circumstances of our current economic existence; a history that continues; a past and a future.

Empire Falling is a series of images *and* a reverberating discourse, an object and a dialogue. As theorist George Baker has written, in an essay on photography and abstraction, "One can only represent the unrepresentable by playing its own game."[7] Dorfman's cunning is to have discerned in something so hard and inert as stone a seismic chord, a polyphony of lively abstract ideas, and to have intuited how to give this a canny visual form. The result is a vision of continuance, of obsolescence refracting toward rebirth, of the past haunting the future. The paradoxes of photography are those of time itself, in which everything and nothing is set in stone.

Endnotes

1 Peter Applebome, "Our Towns: Cool Waters, and the Lure of Danger," *The New York Times*, July 2, 2009.

2 For example: Dan Luzadder, "San Diego Reinvents a Fading Quarry," *The New York Times*, Mar. 4, 2009.

3 Charles H. Richardson, *The Building Stones of Kentucky* (Lexington: The Kentucky Geological Survey, 1923).

4 Dale Gnidovec, paraphrased in, "Field Tripping in the Pits: CLUI Leads Ohio State into Aggregate," *The Center for Land Use Interpretation Newsletter* (Winter 2012).

5 A recent summation: Mia Fineman, *Faking It: Manipulated Photography Before Photoshop*, exh. cat. (New York: Metropolitan Museum of Art, 2012).

6 George Baker explores these ideas, borrowing terms from Frederic Jameson, in his essay, "Photography and Abstraction," in *Words Without Pictures*, eds. Charlotte Cotton and Alex Klein (Los Angeles: LACMA/ Aperture, 2009).

7 *Ibid*, 370.

Empire Falling 2, 2011, 37.5 x 50 inches, Detail pp. 14 - 15

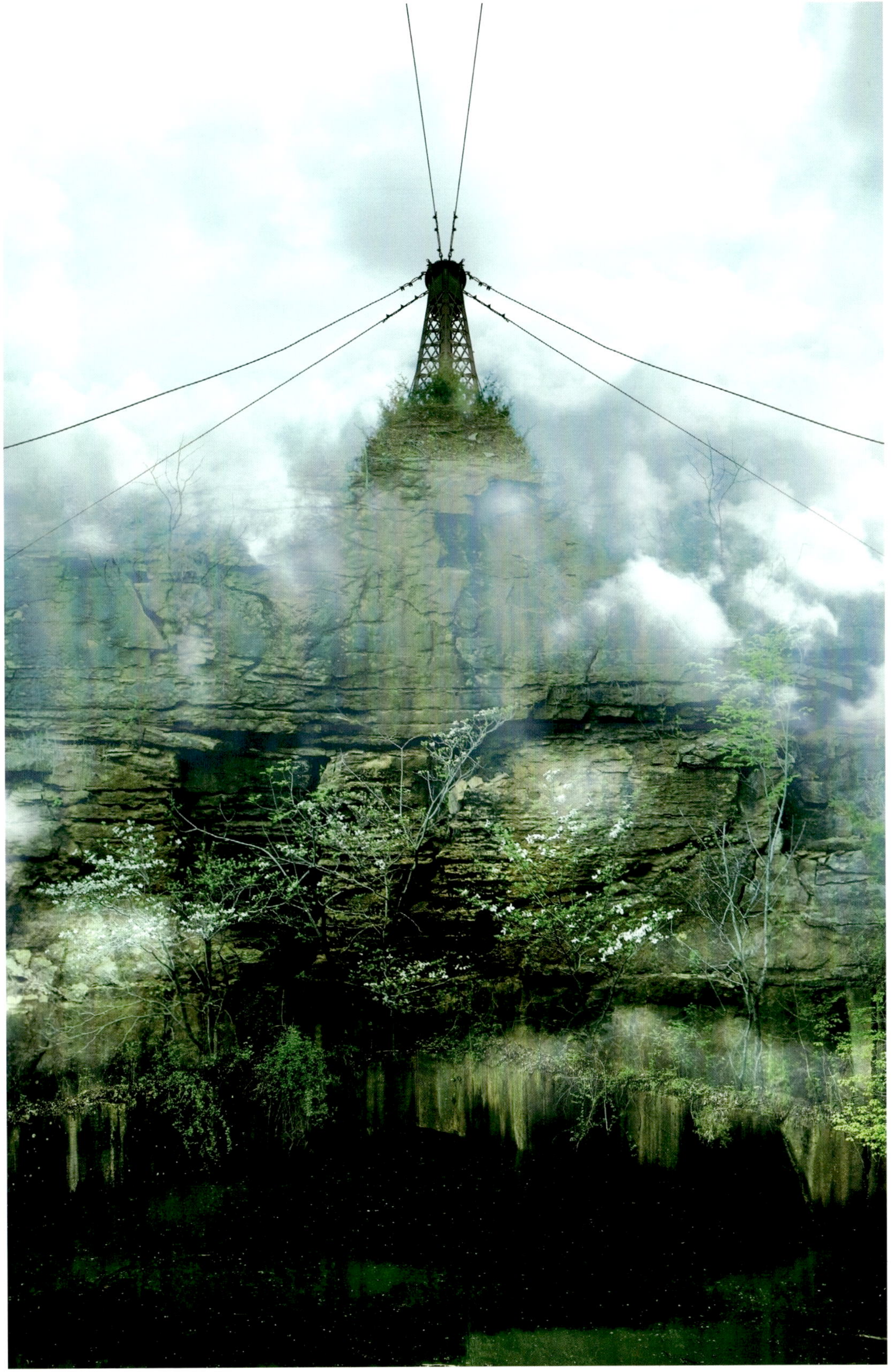

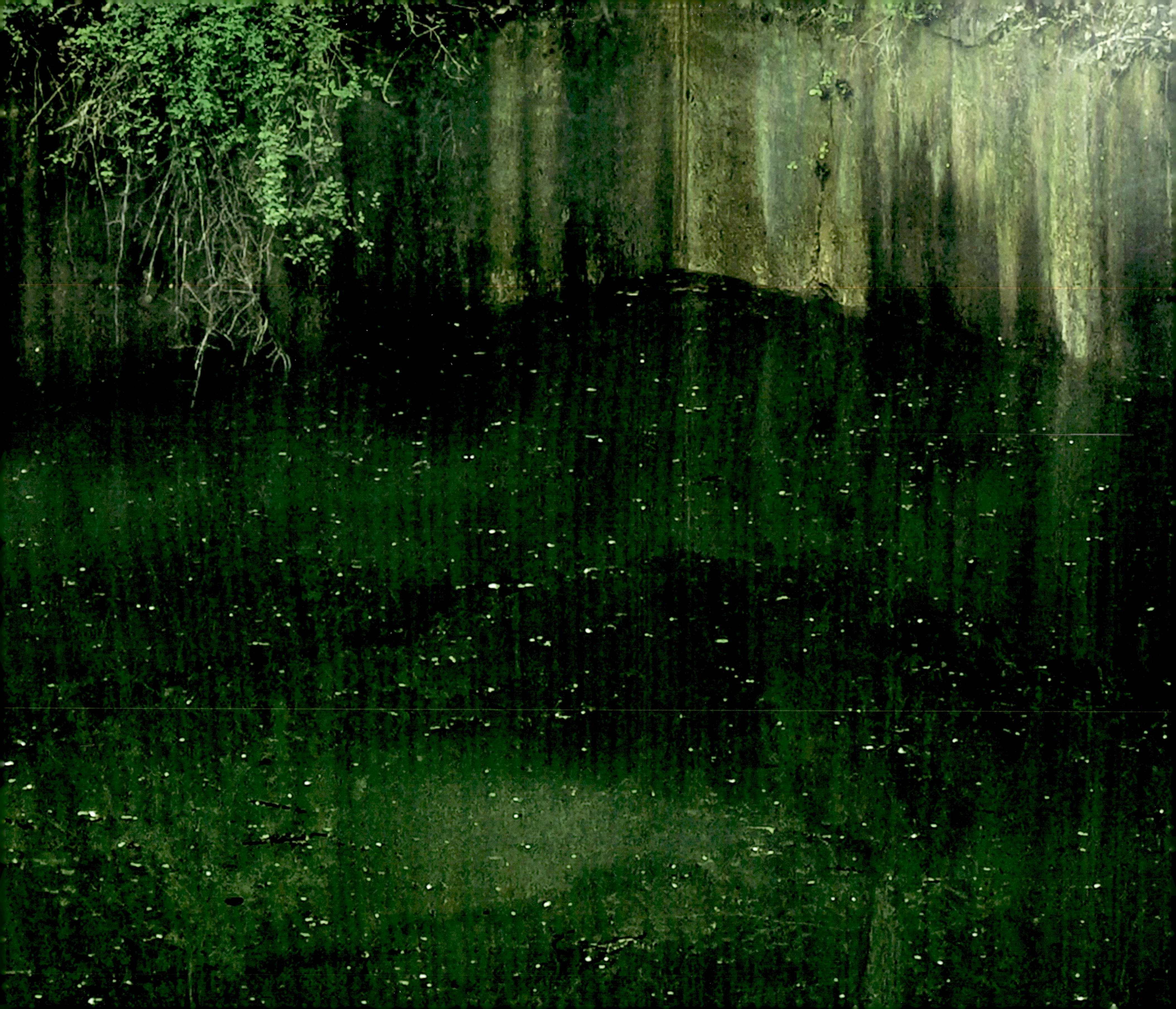

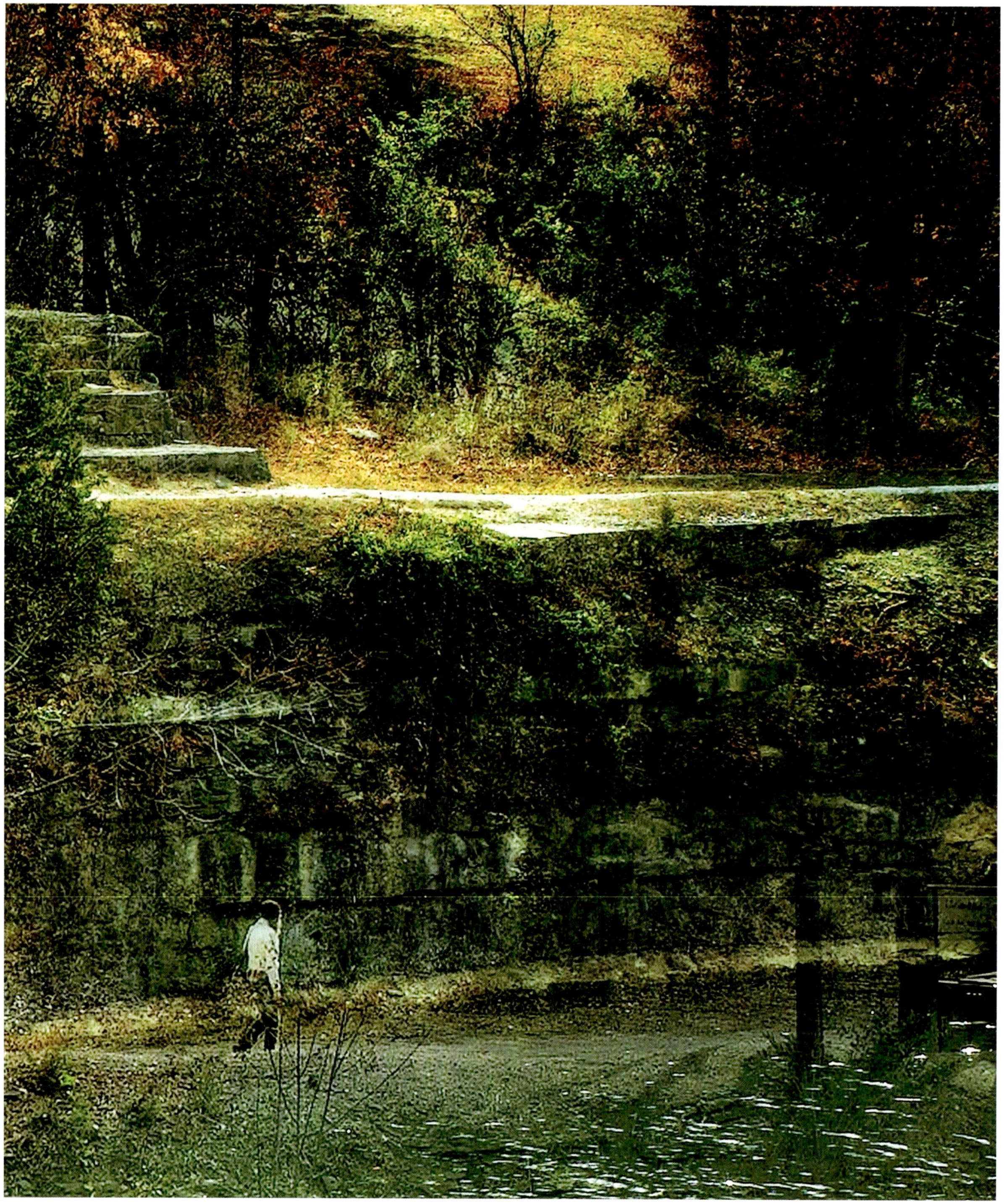

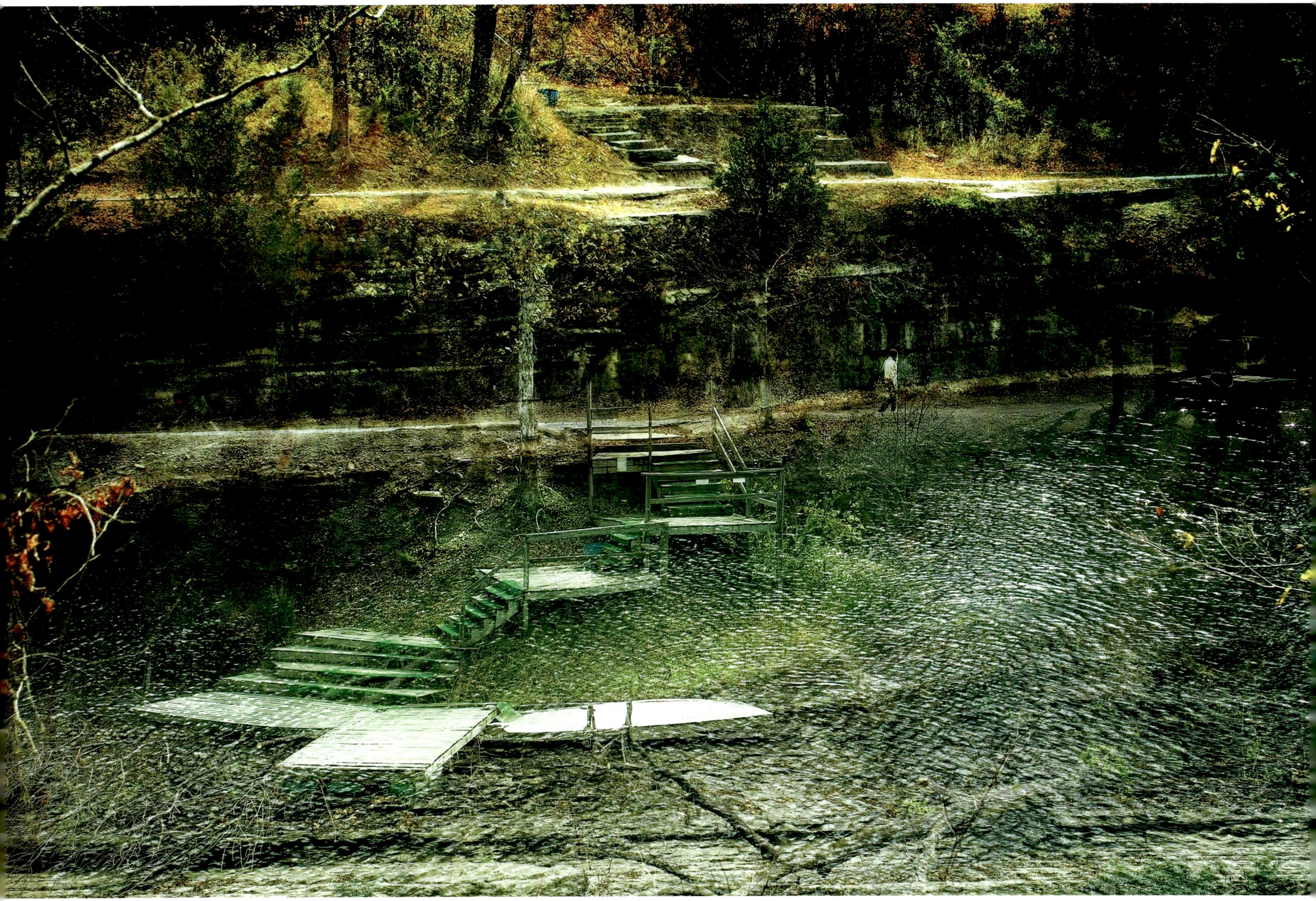

Empire Falling 3, 2011, 37.3 x 50 inches, Detail p. 17

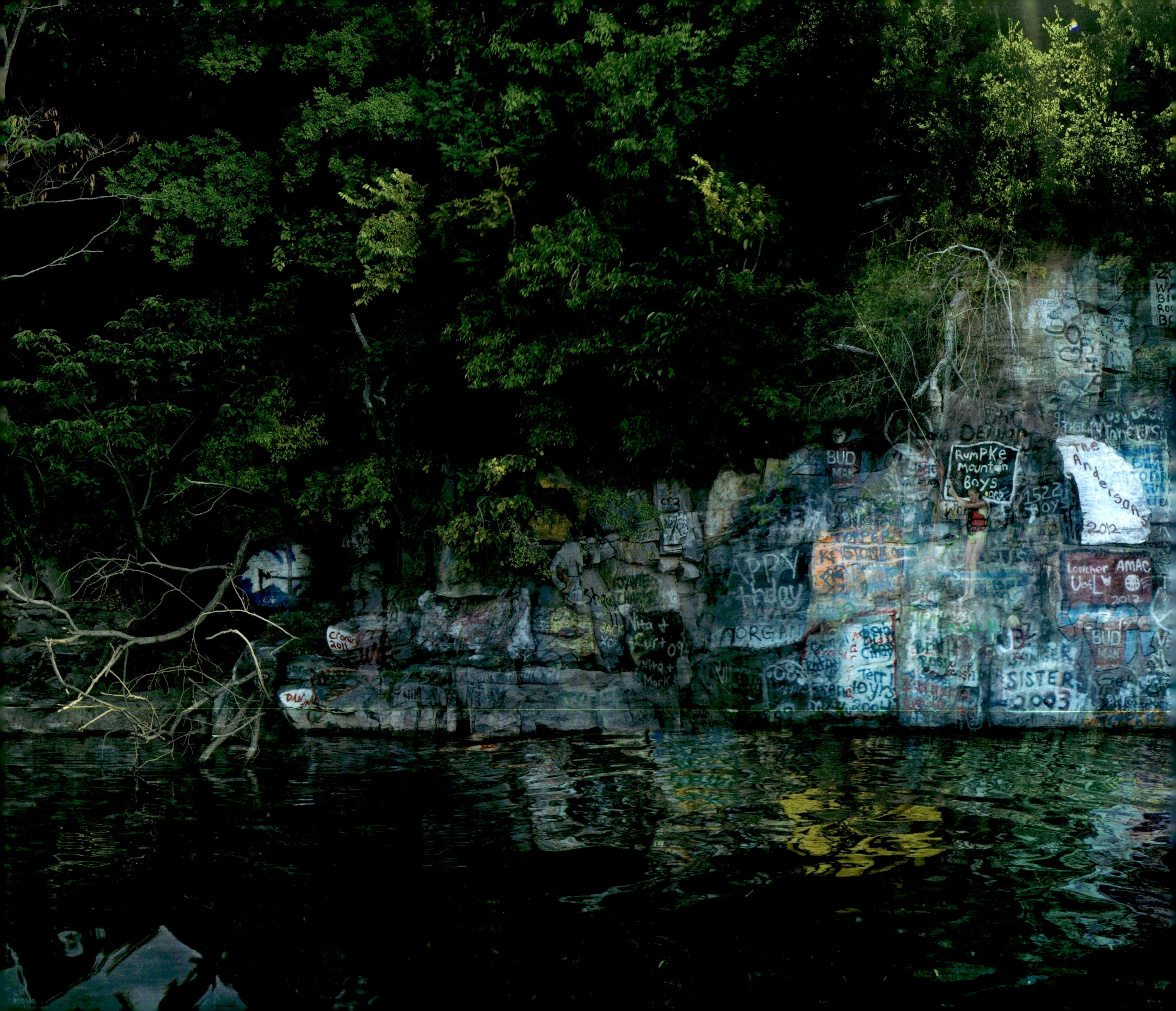
RumPKe Mountain Boys
The Andersons
2012
BUD MAN
SISTER 2003
MORGAN

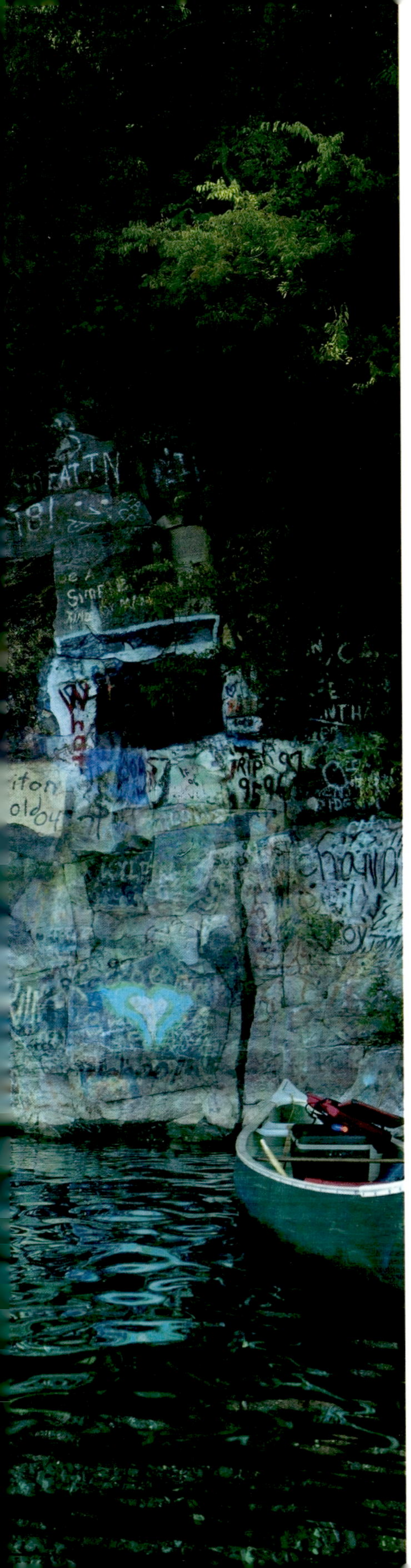

Empire Falling 4, 2012, 33.3 x 50 inches, Detail pp. 20 - 21

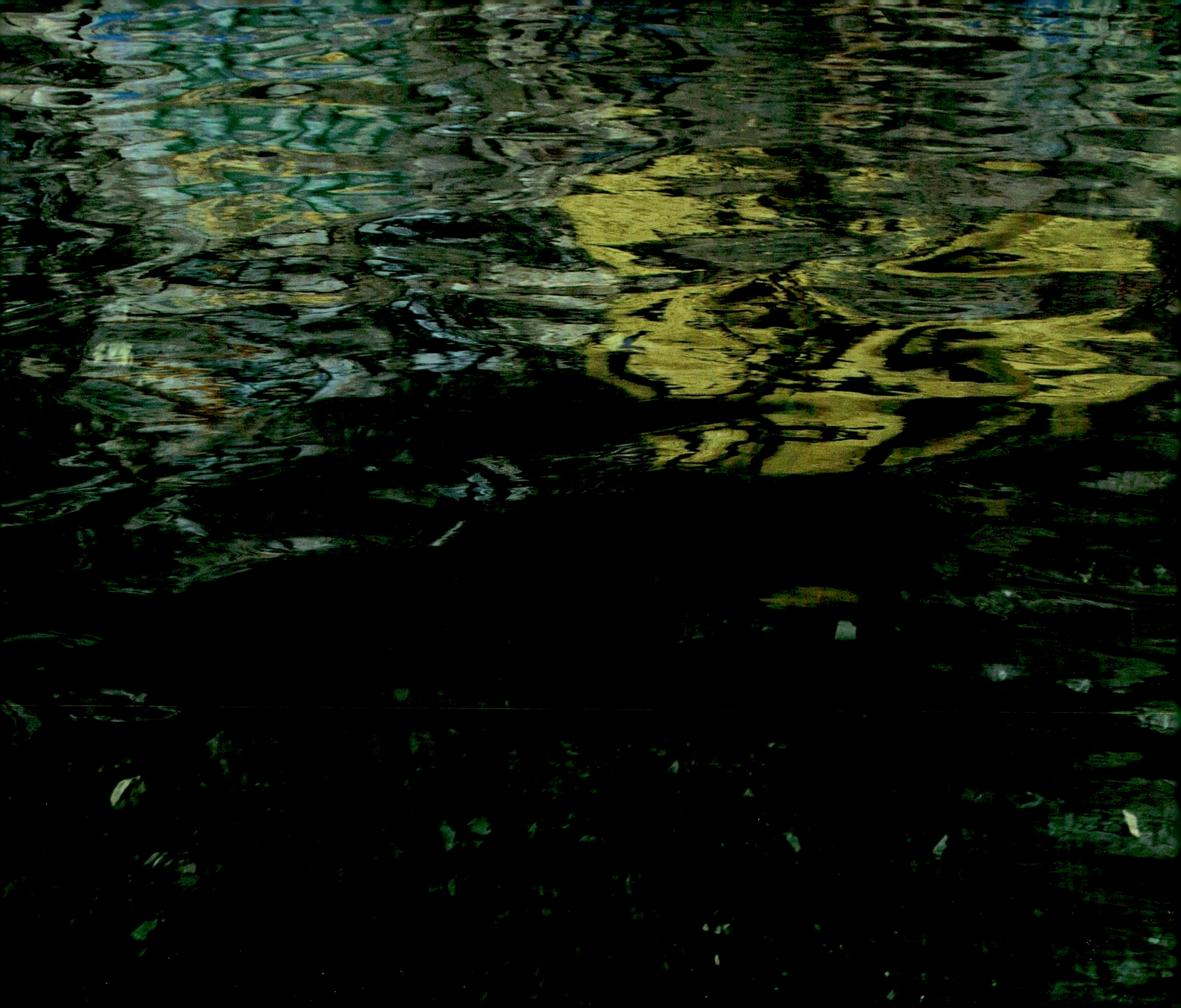

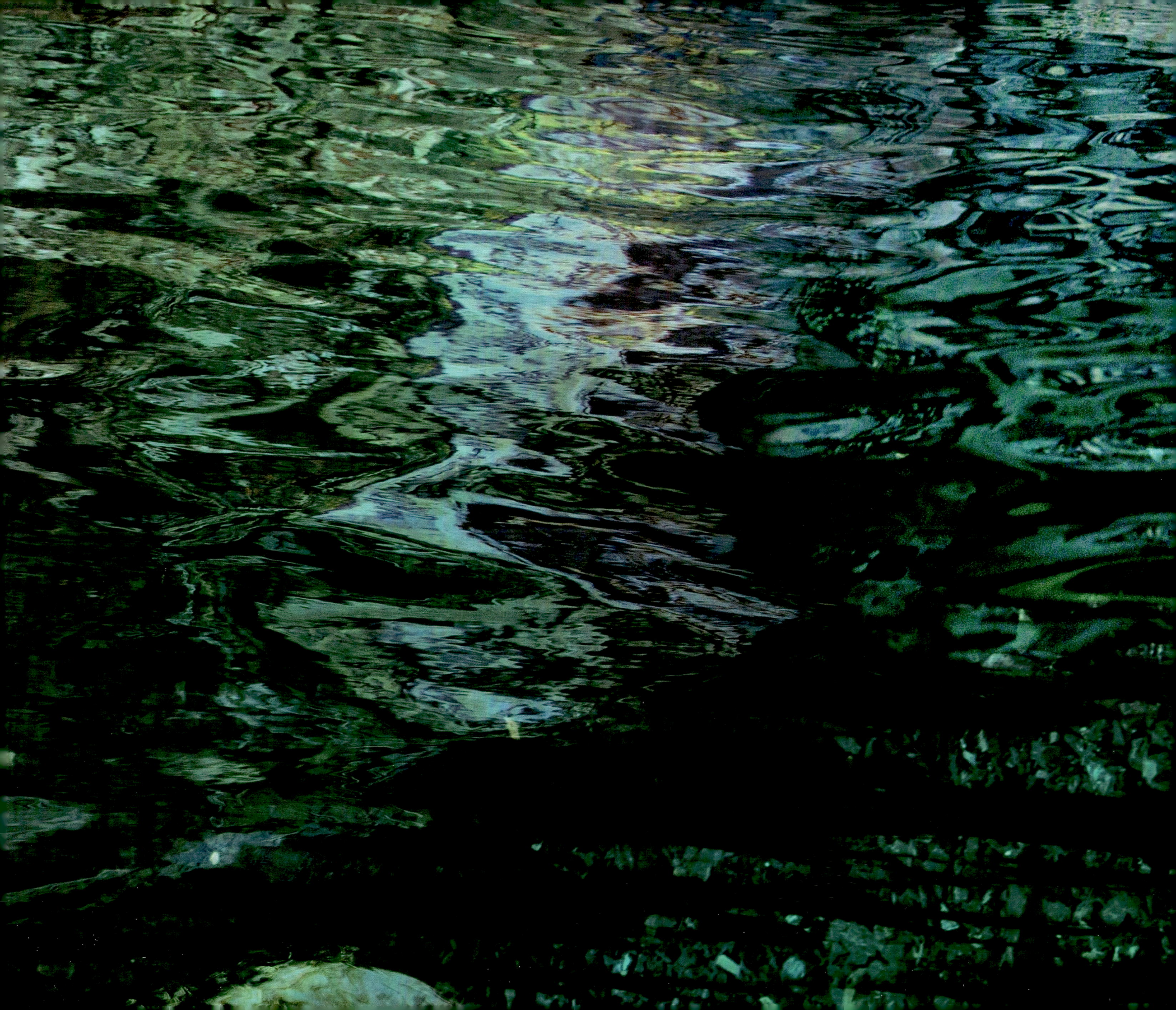

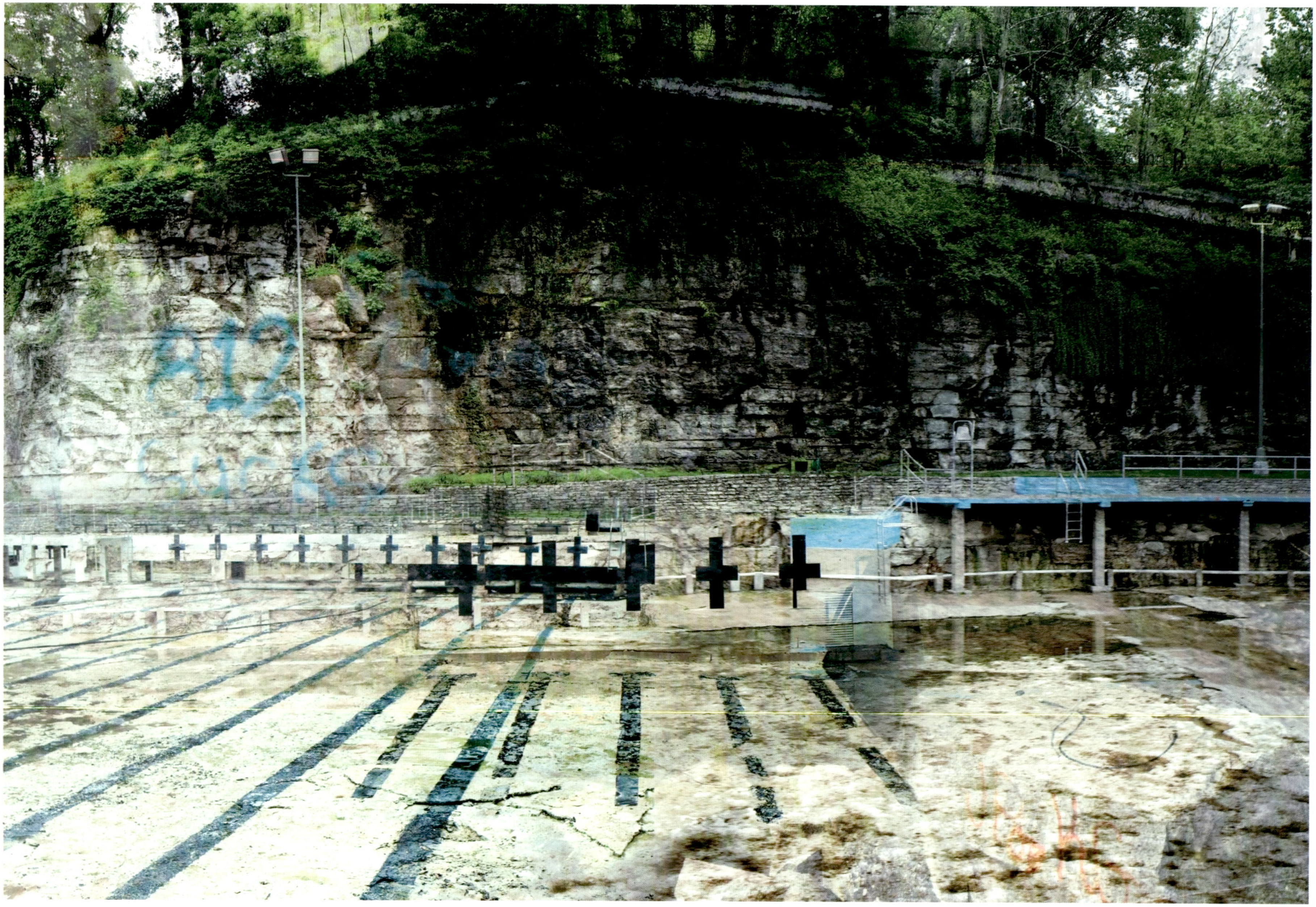

Empire Falling 5, 2011, 33.3 x 50 inches, Detail p. 23

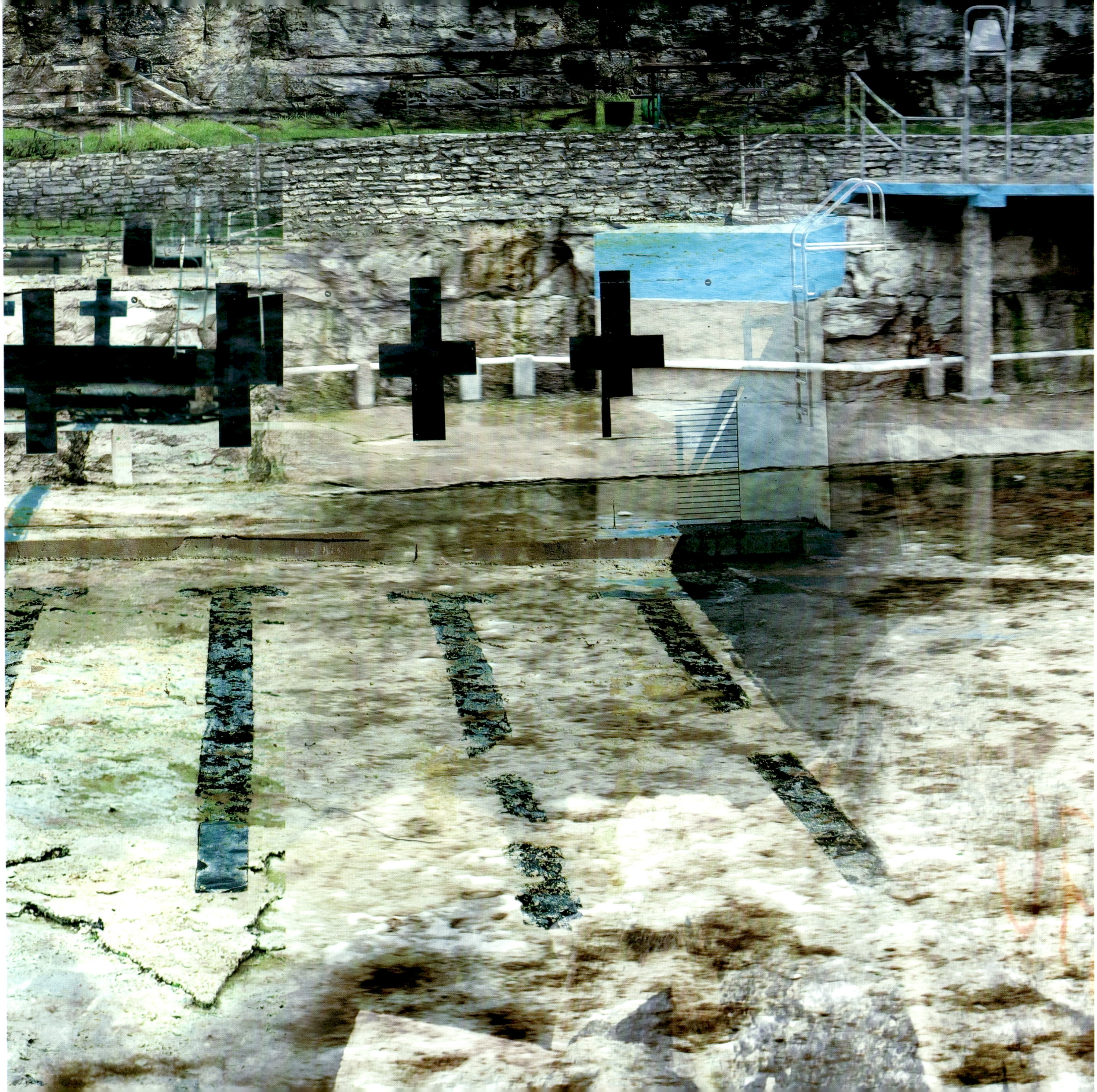

Empire Falling 7, 2011, 33.3 x 50 inches, Detail pp. 26 - 27

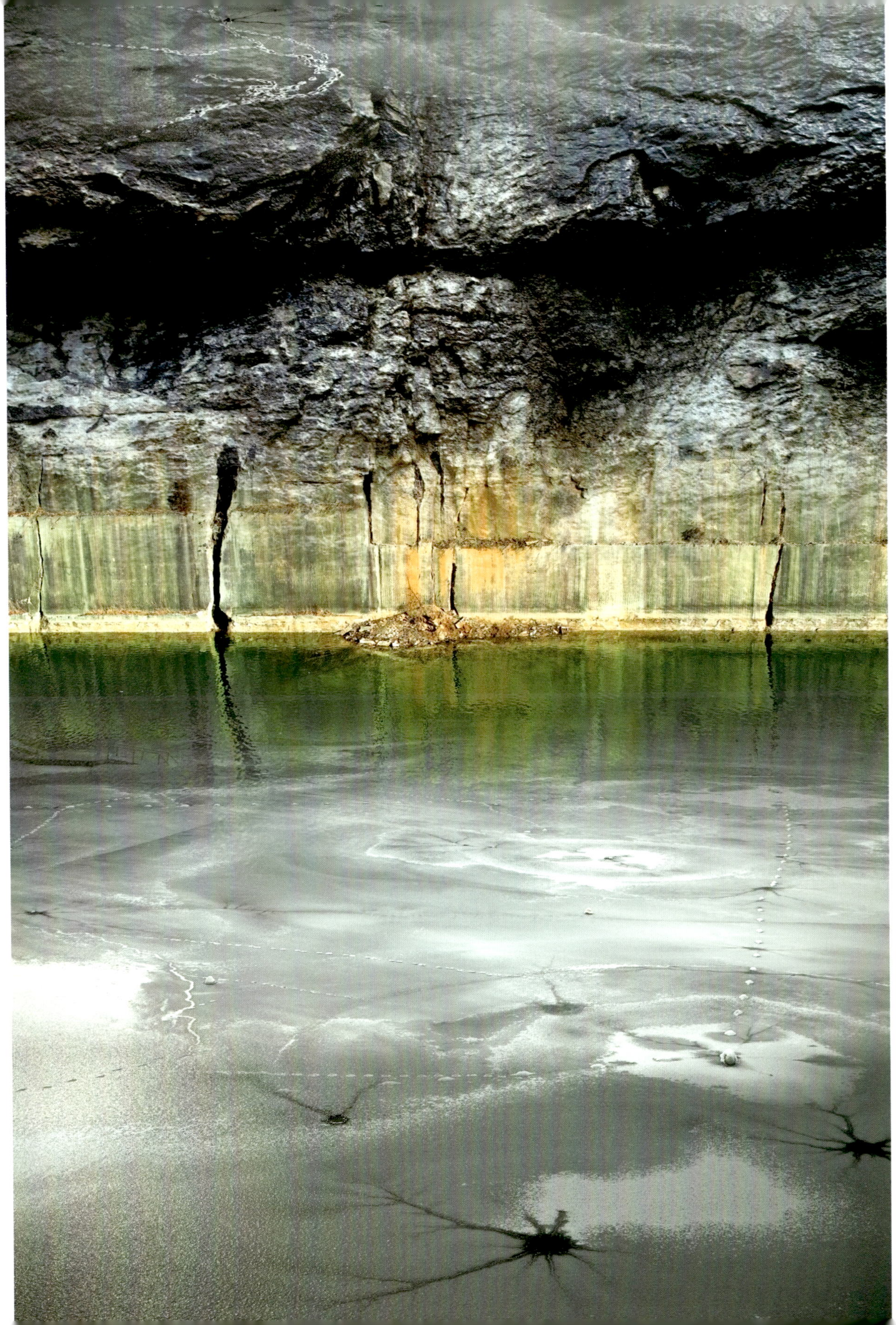

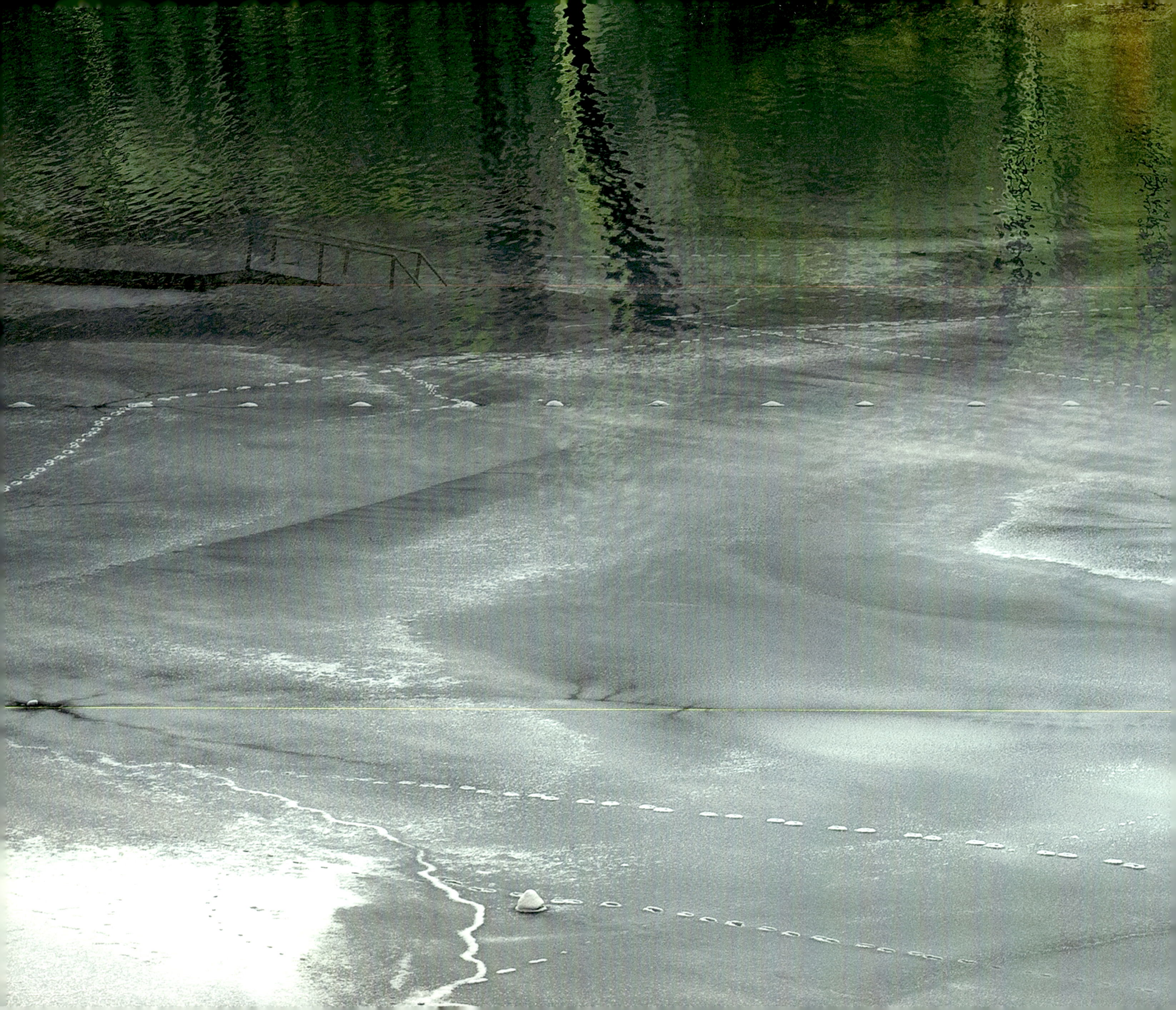

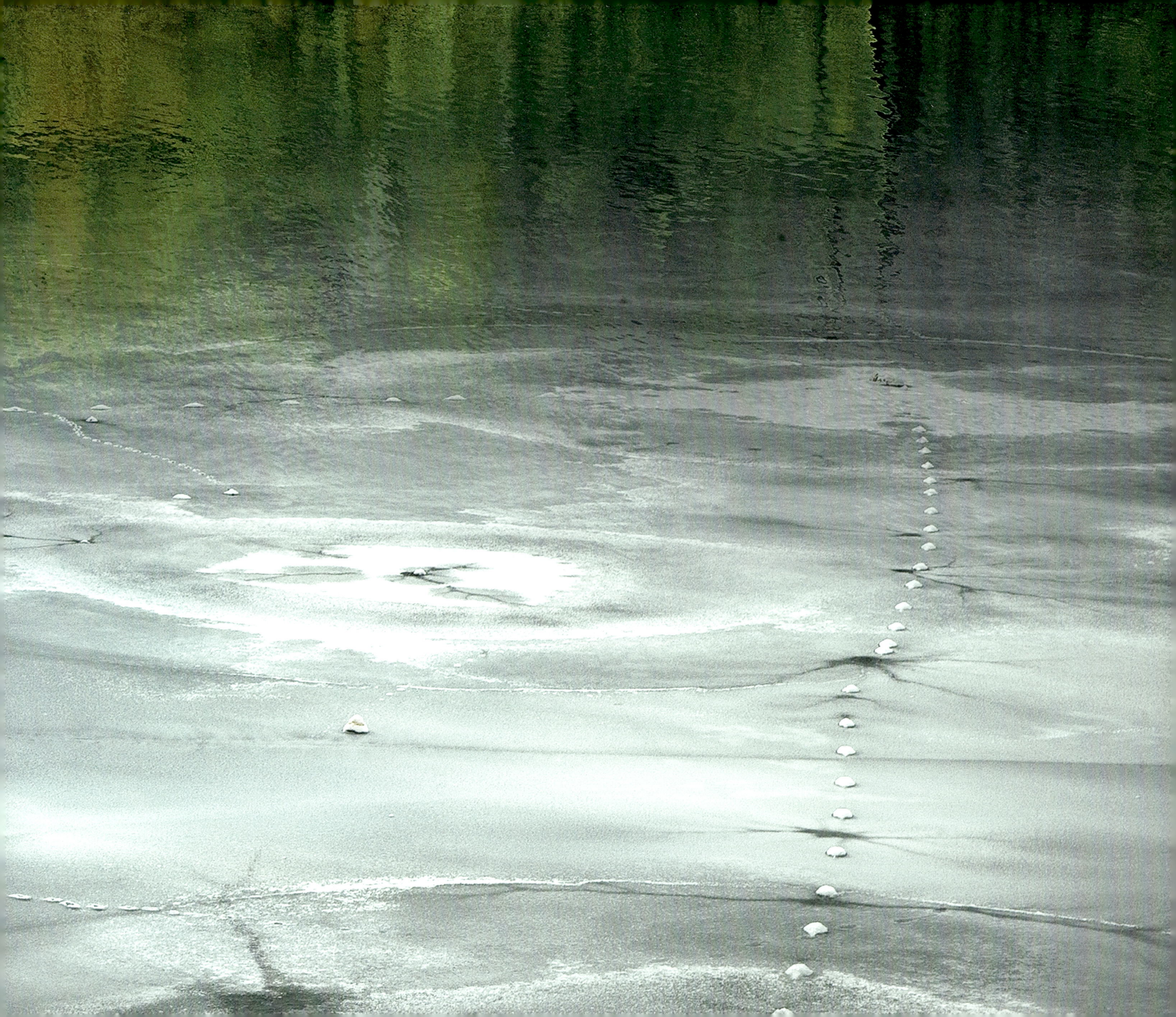

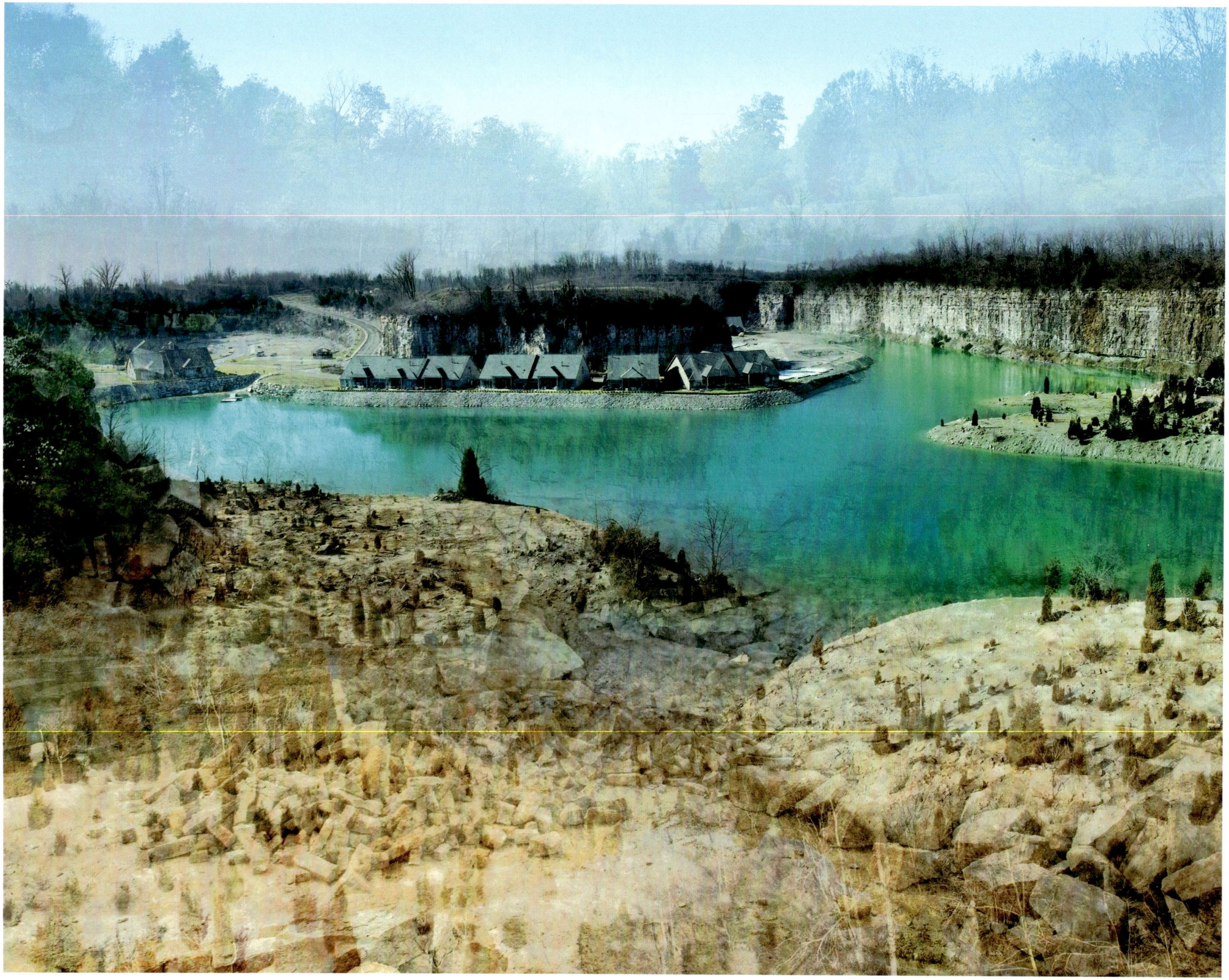

Empire Falling 8, 2012, 37.5 x 50 inches, Details p. 29

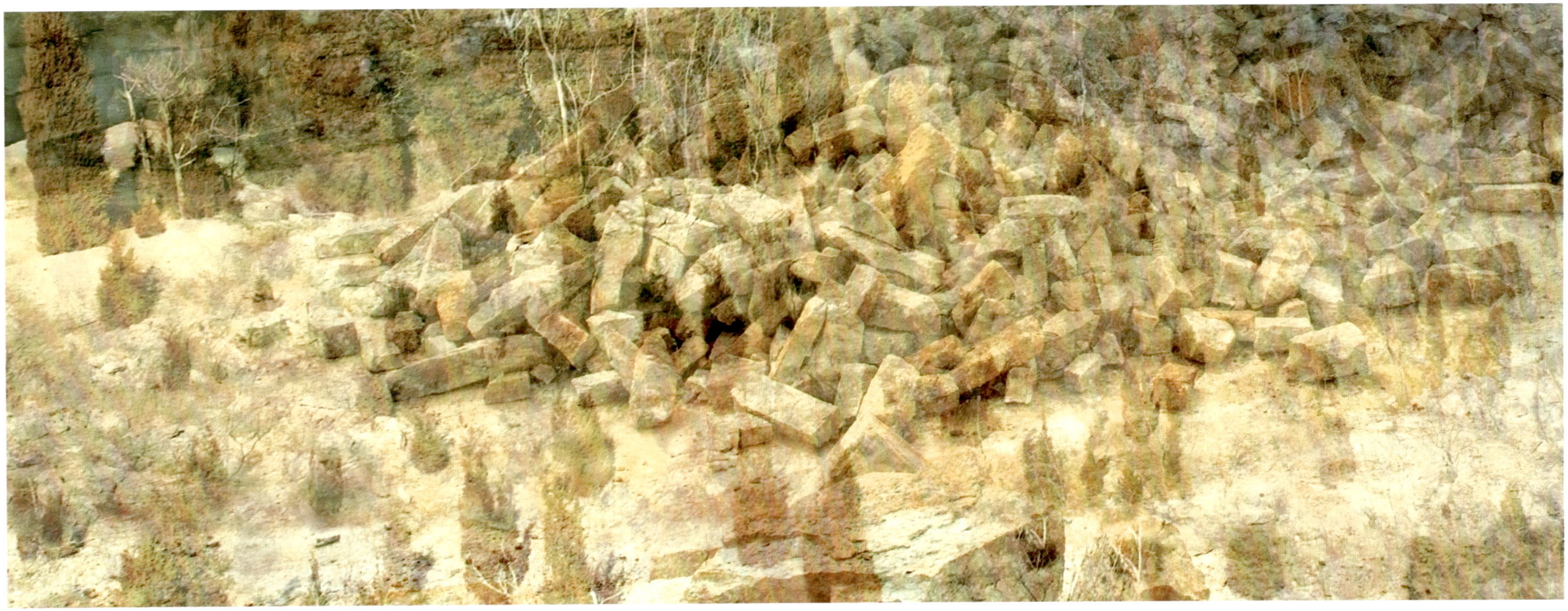

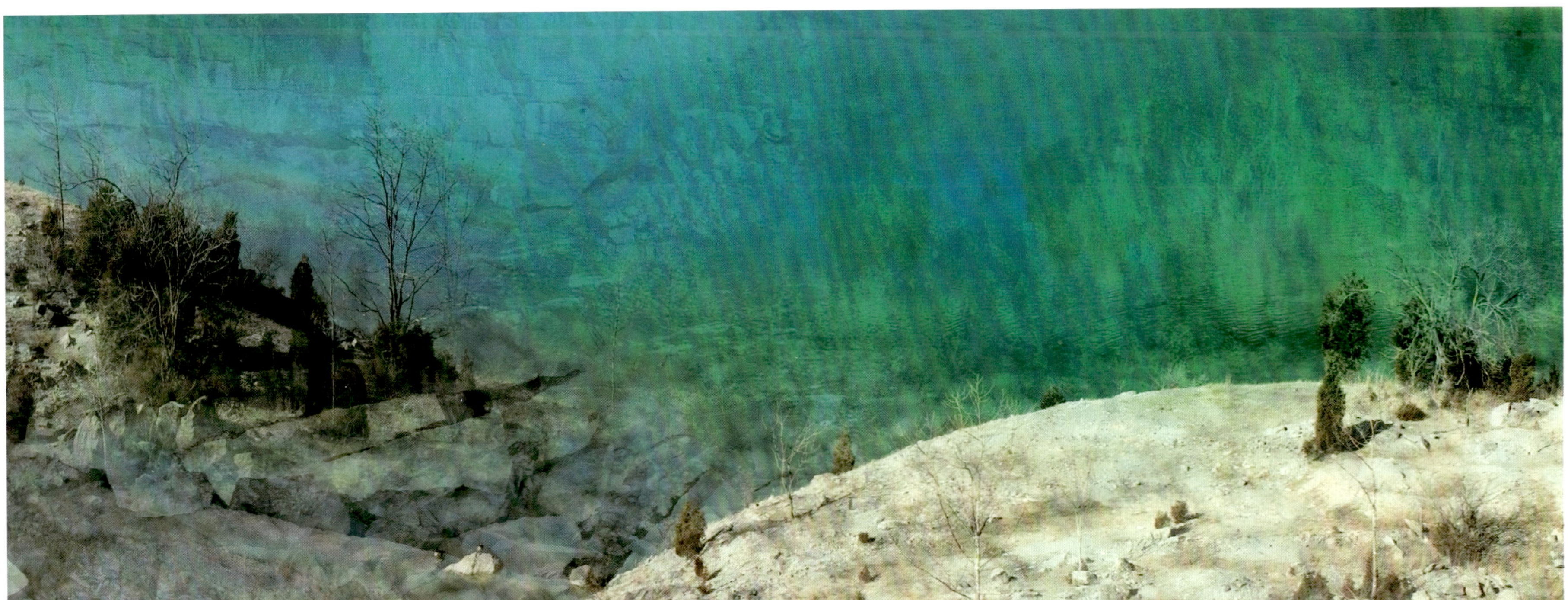

Empire Falling 9, 2012, 37.5 x 50 inches, Detail pp. 32 - 33

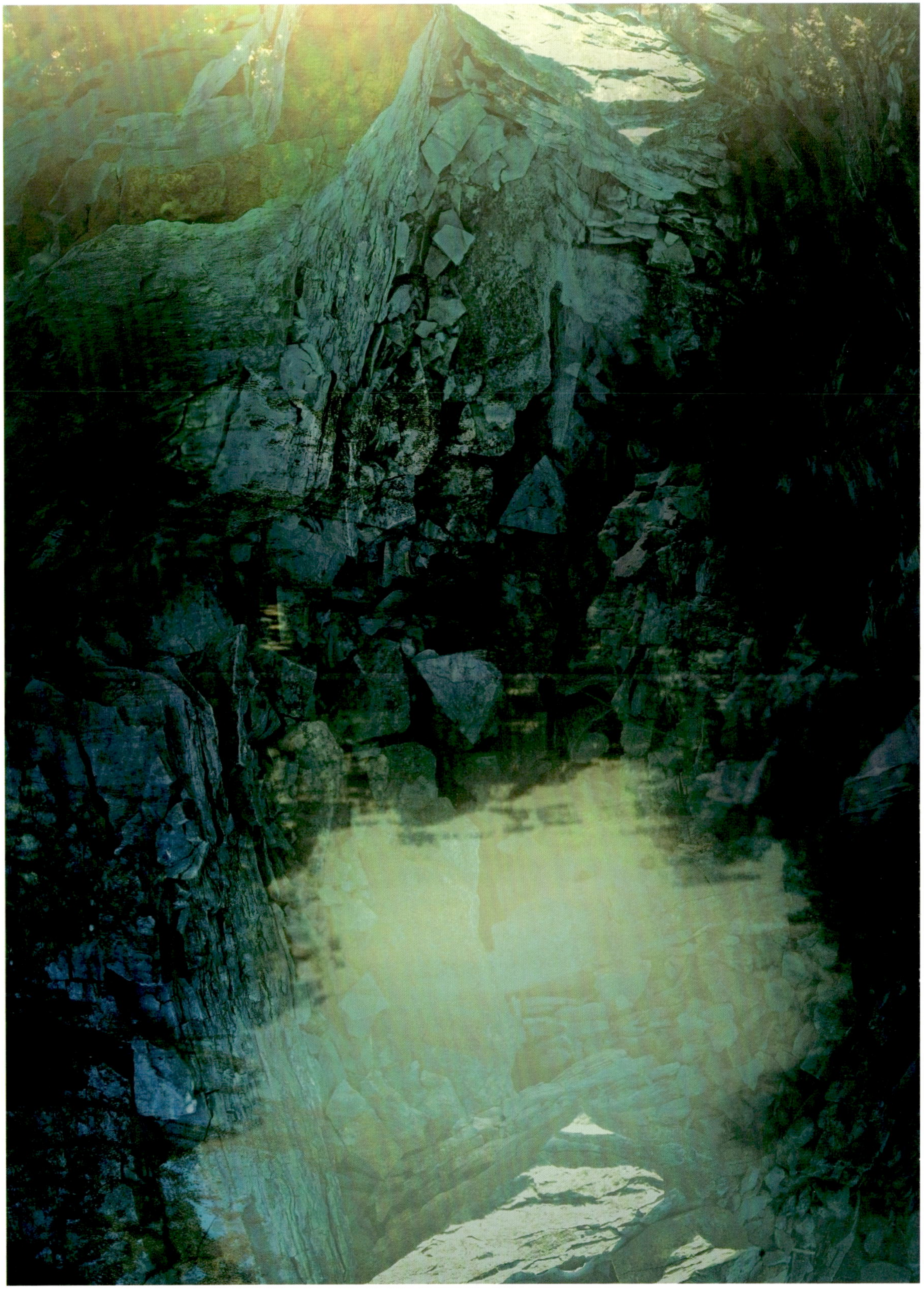

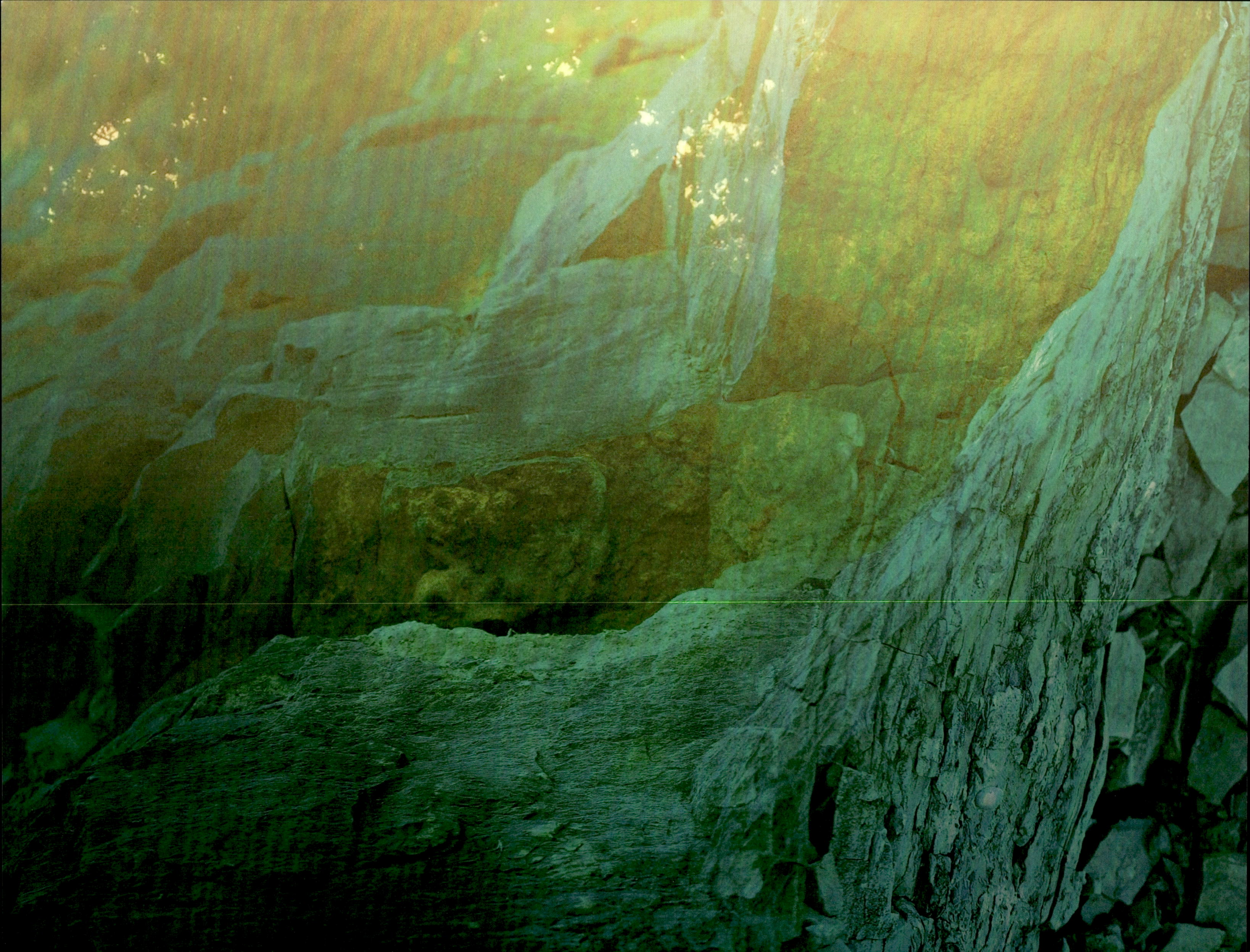

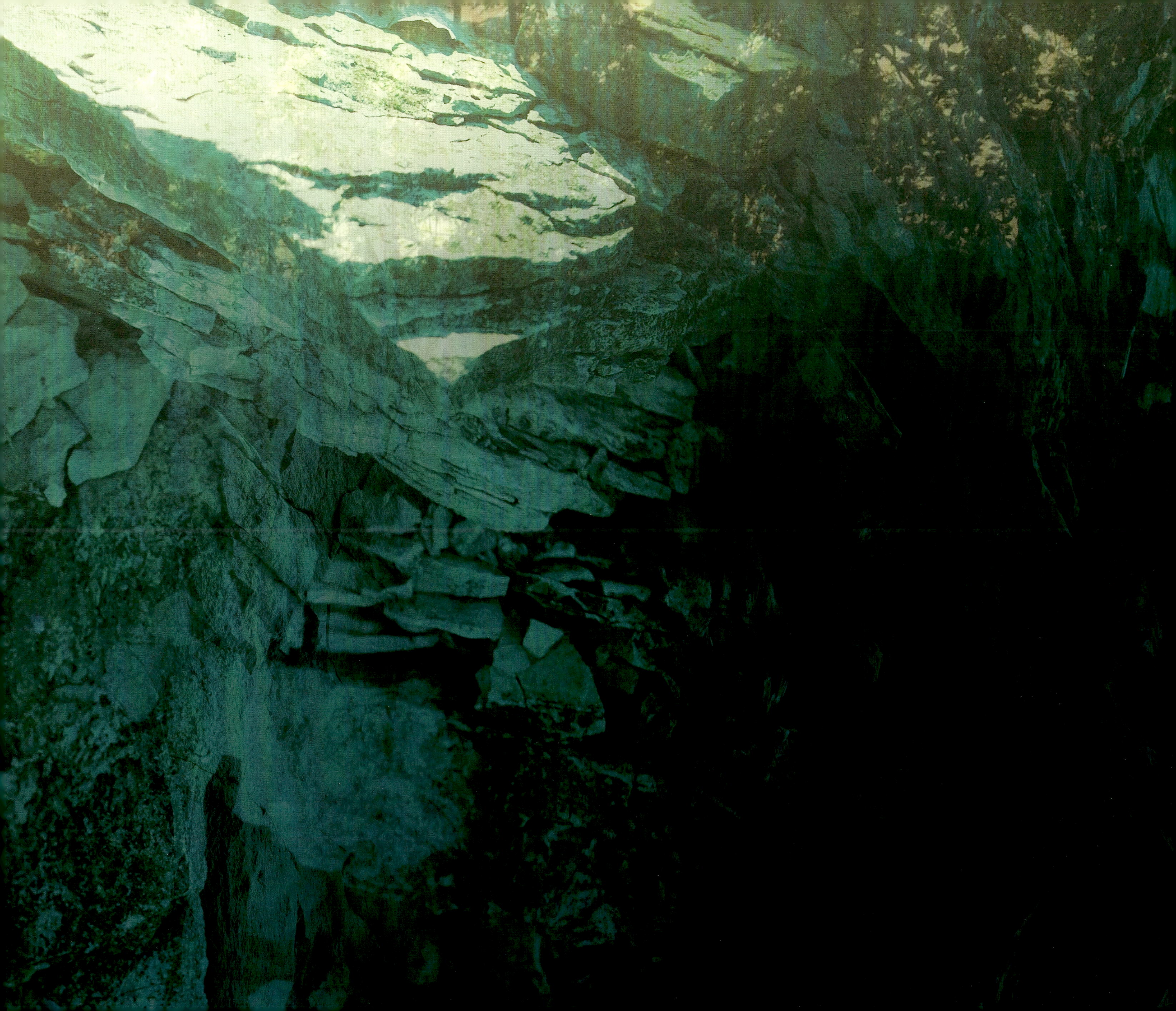

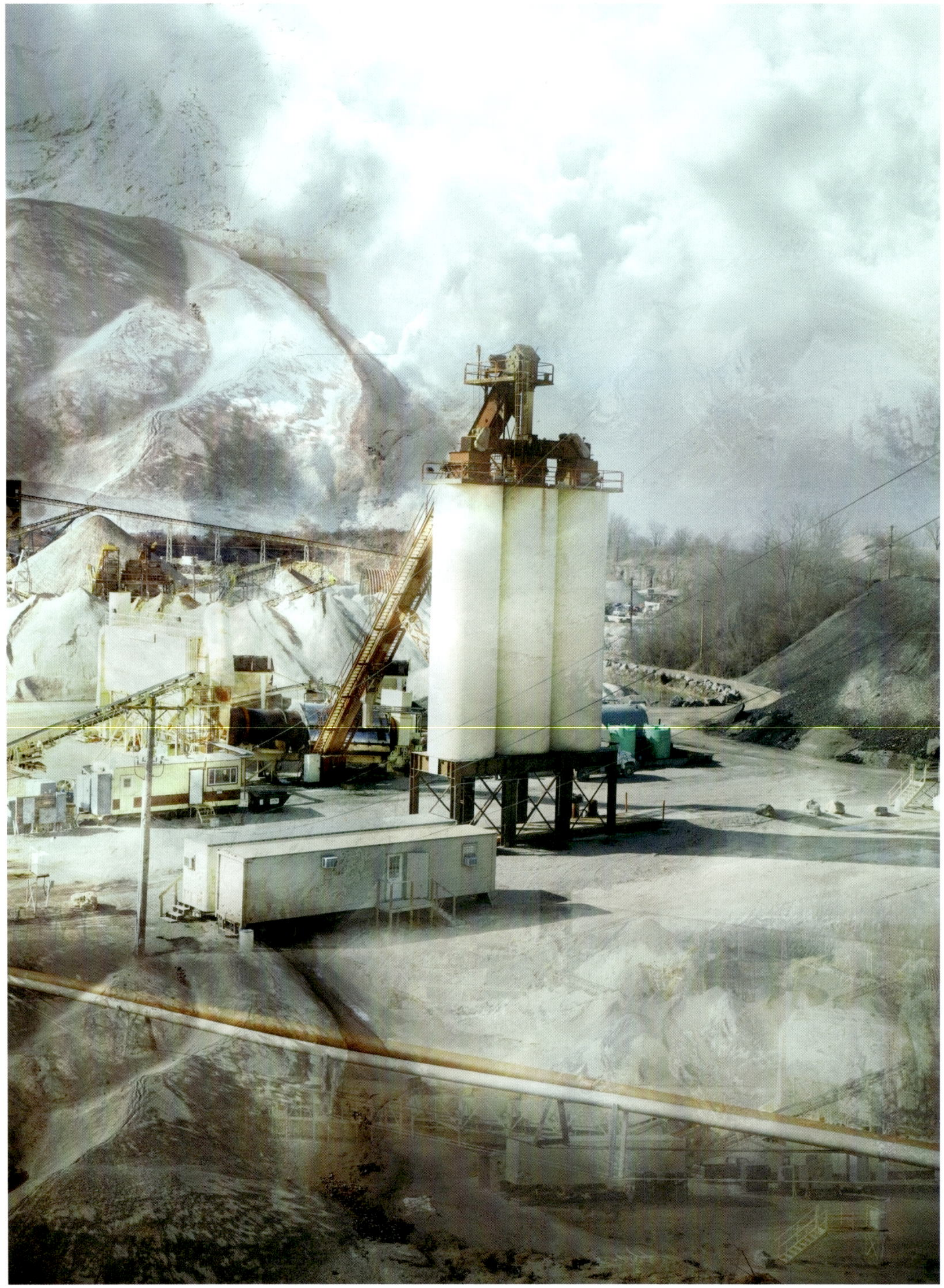

Empire Falling 10, 2011, 37.5 x 50 inches, Detail p. 35

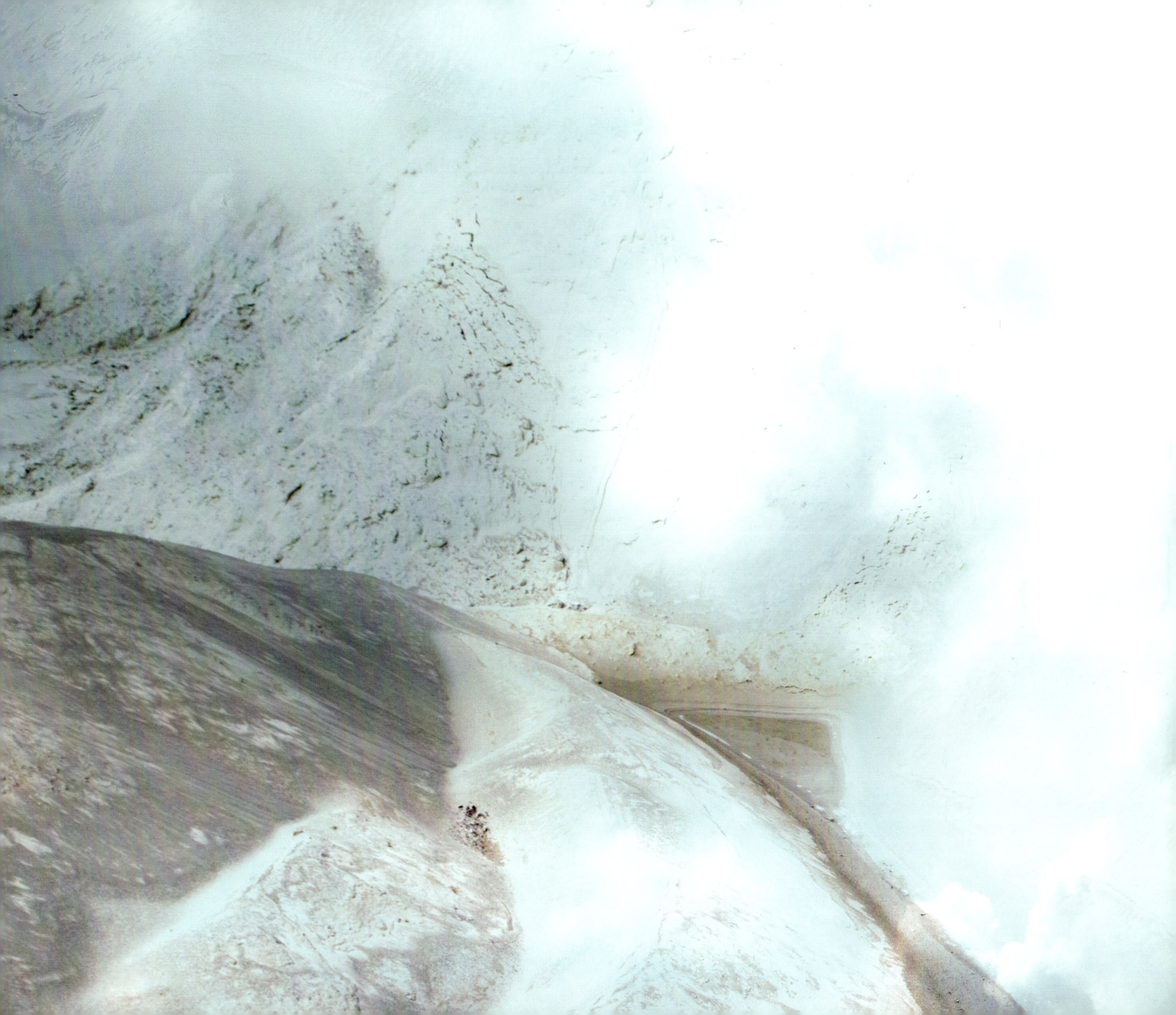

Empire Falling 11, 2011, 37.5 x 50 inches

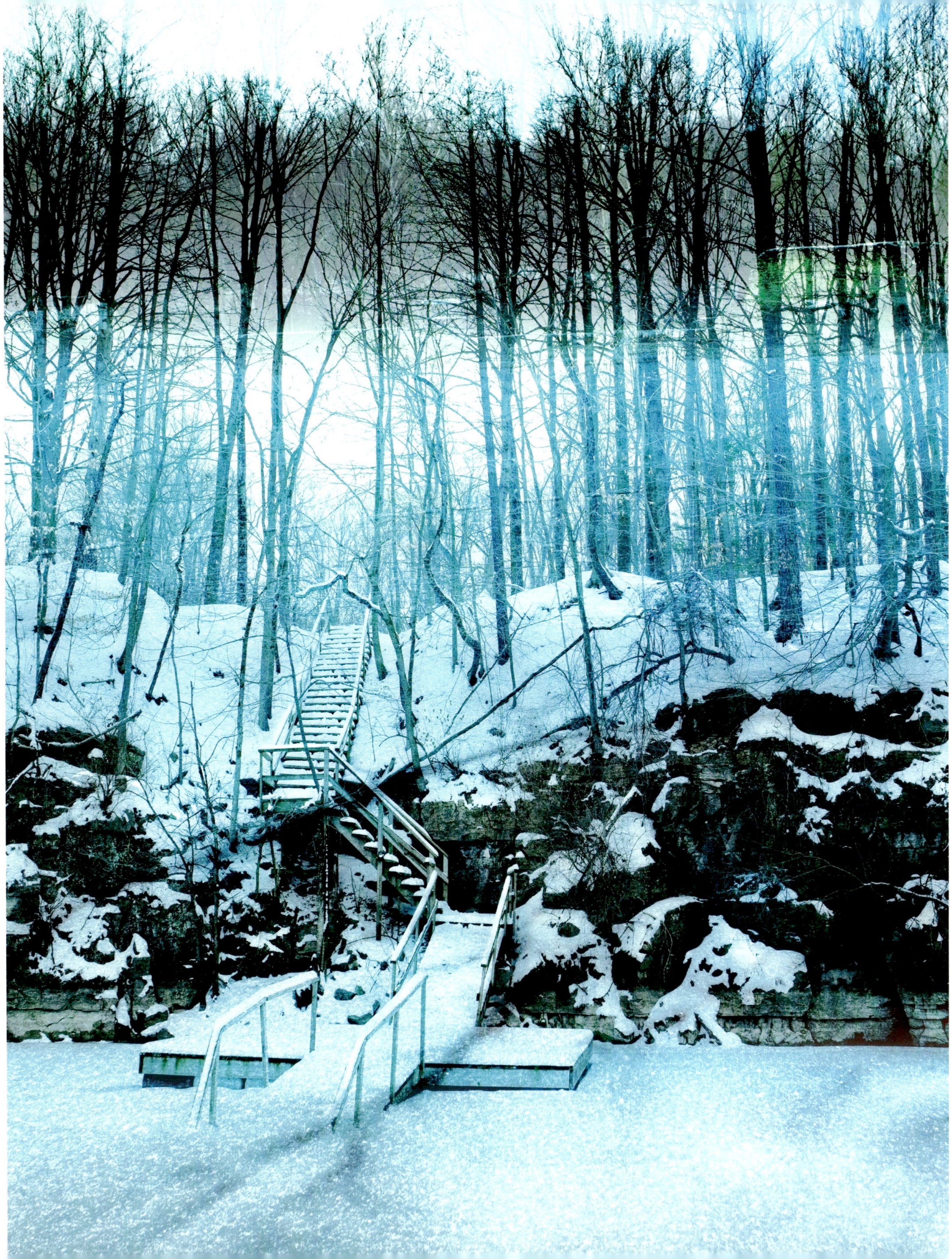

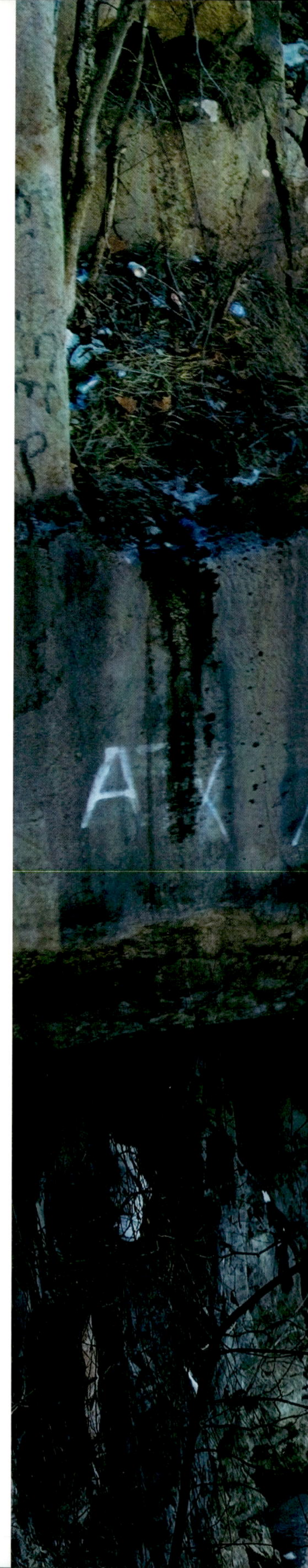

Empire Falling 12, 2012, 37.5 x 50 inches, Detail pp. 40 - 41

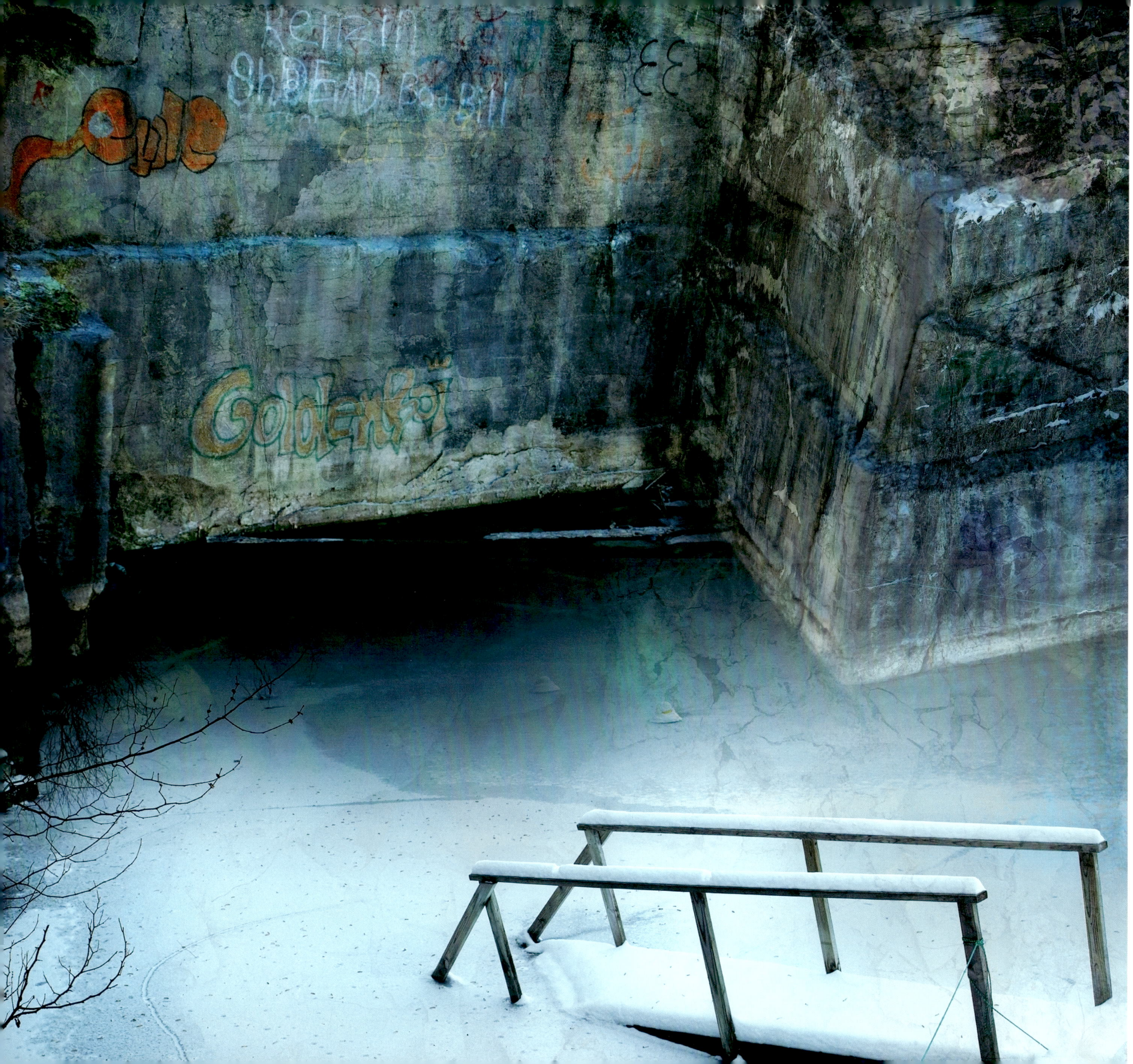

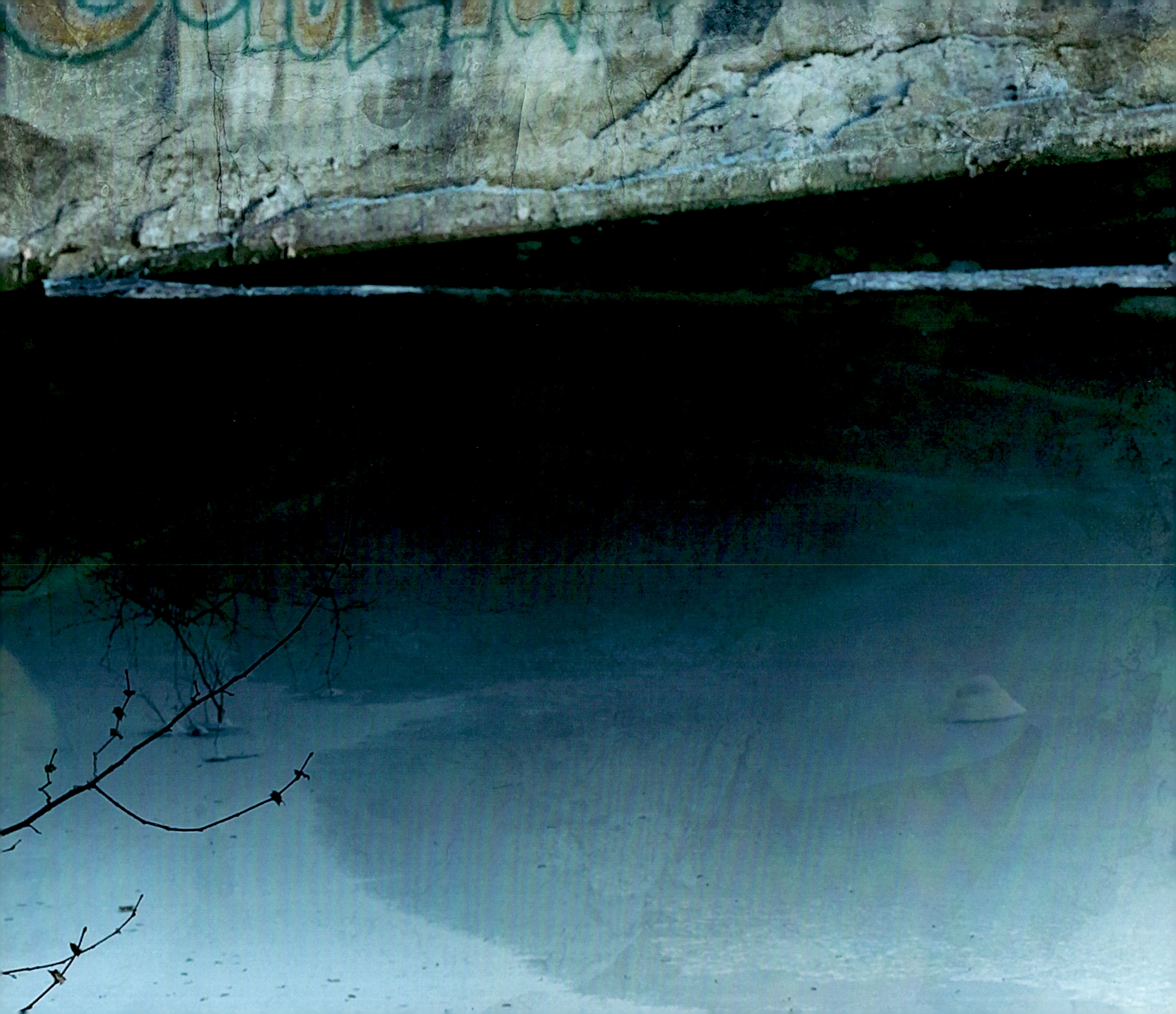

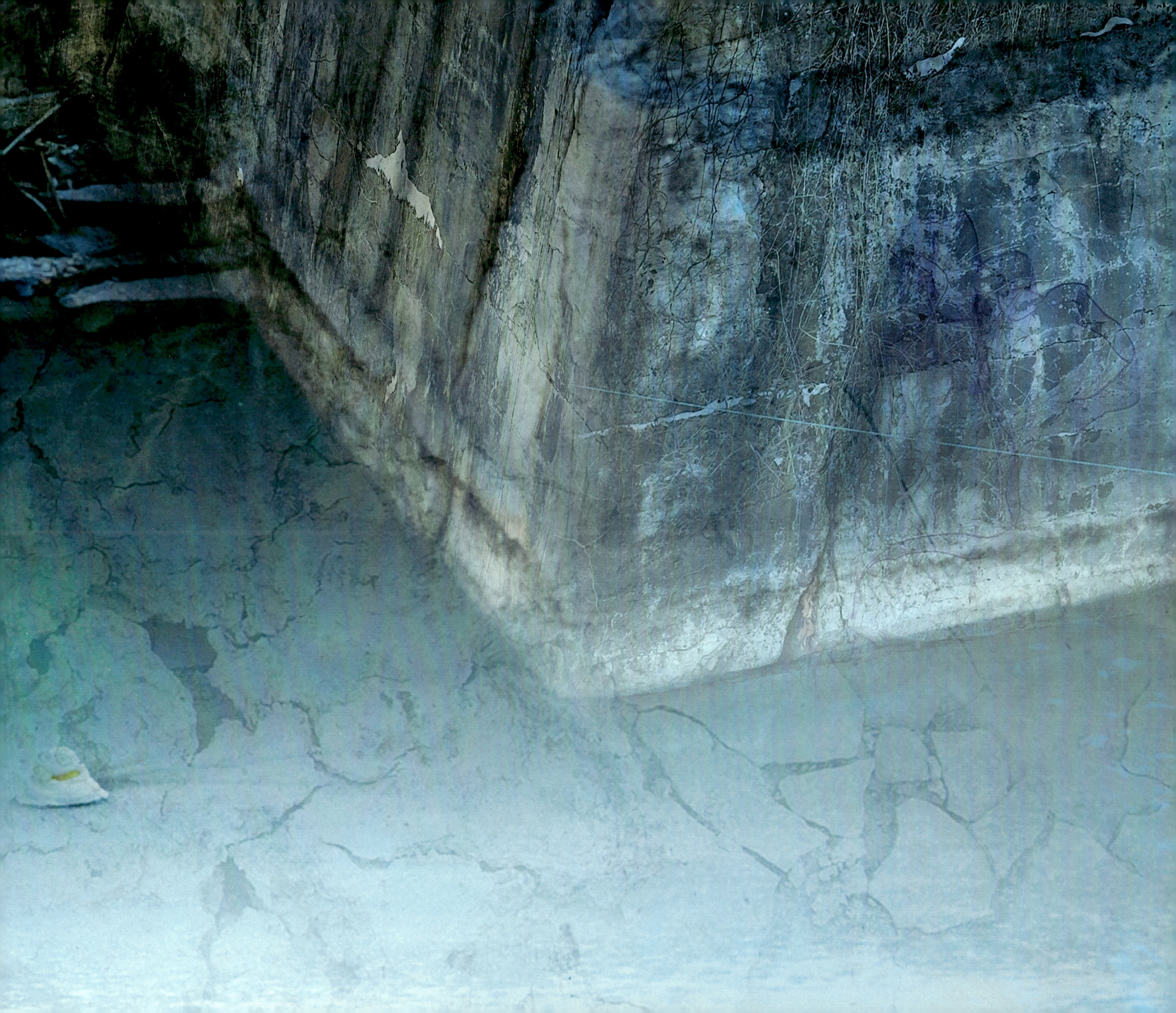

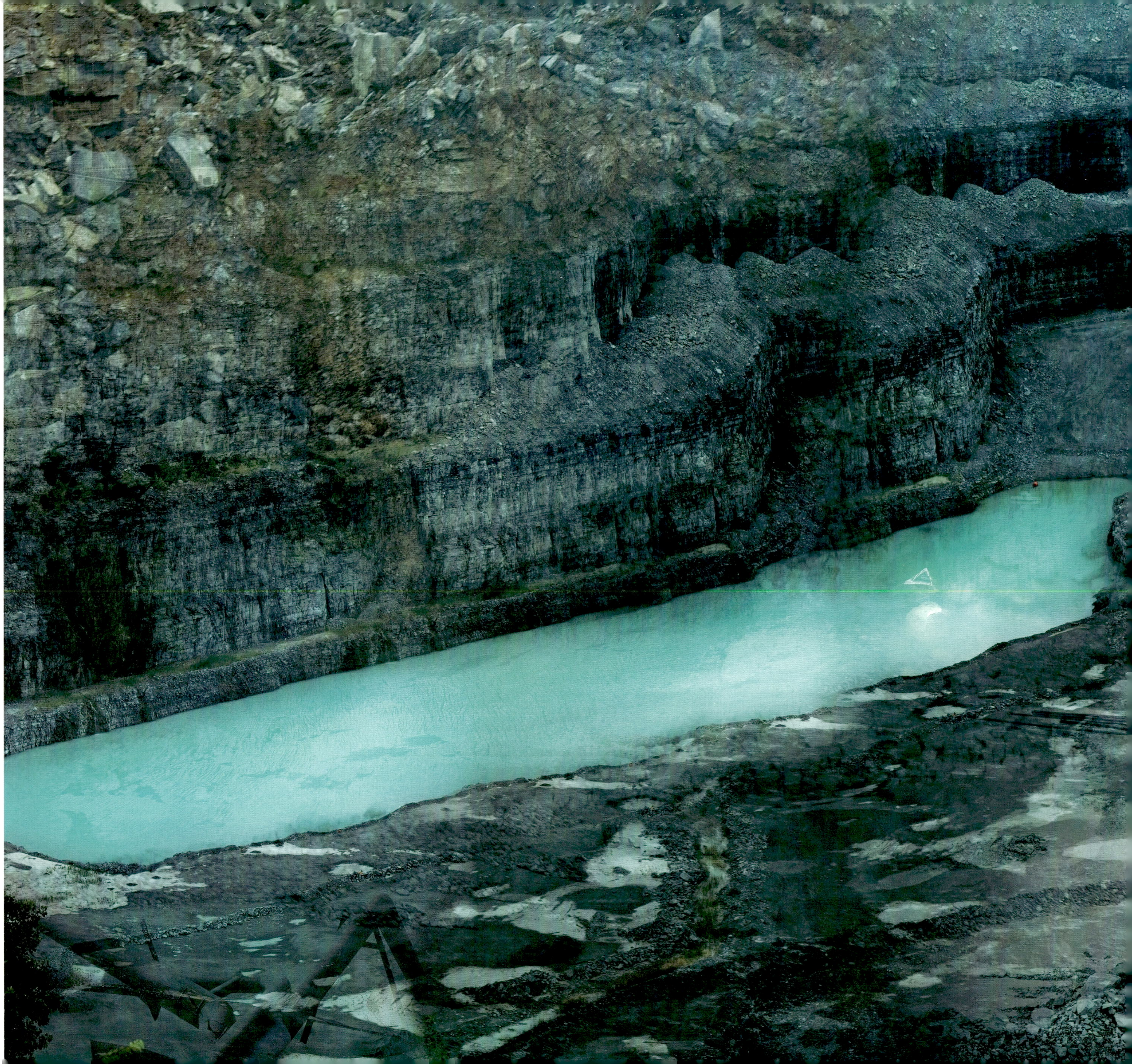

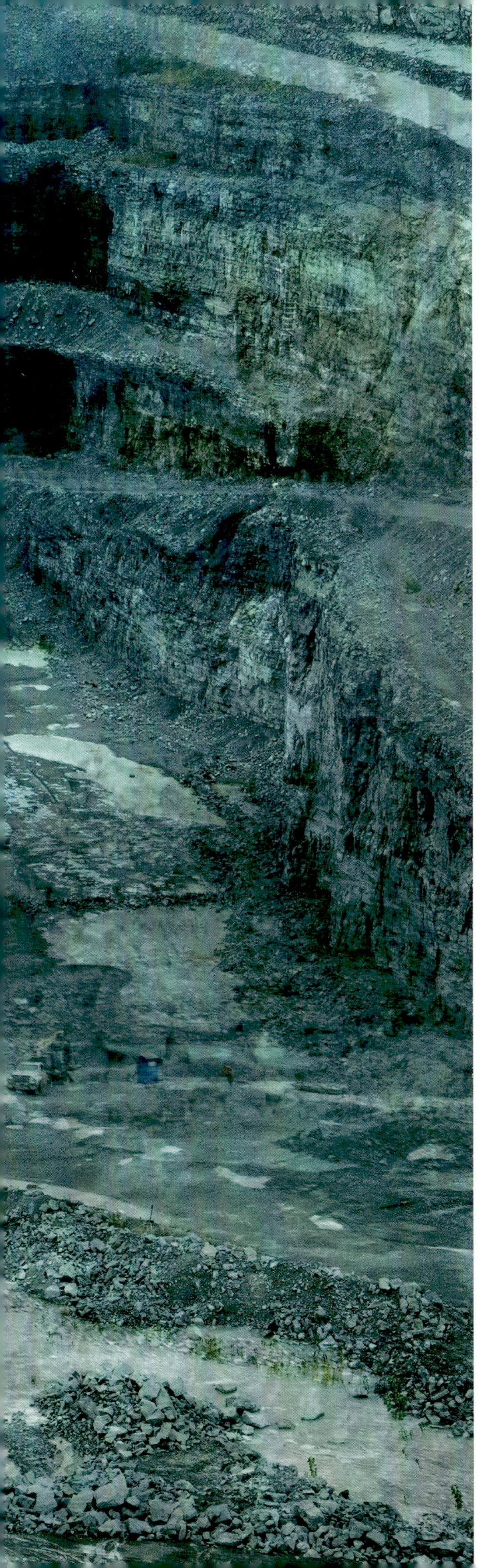

Empire Falling 13, 2012, 37.5 x 50 inches, Details pp. 44 - 47

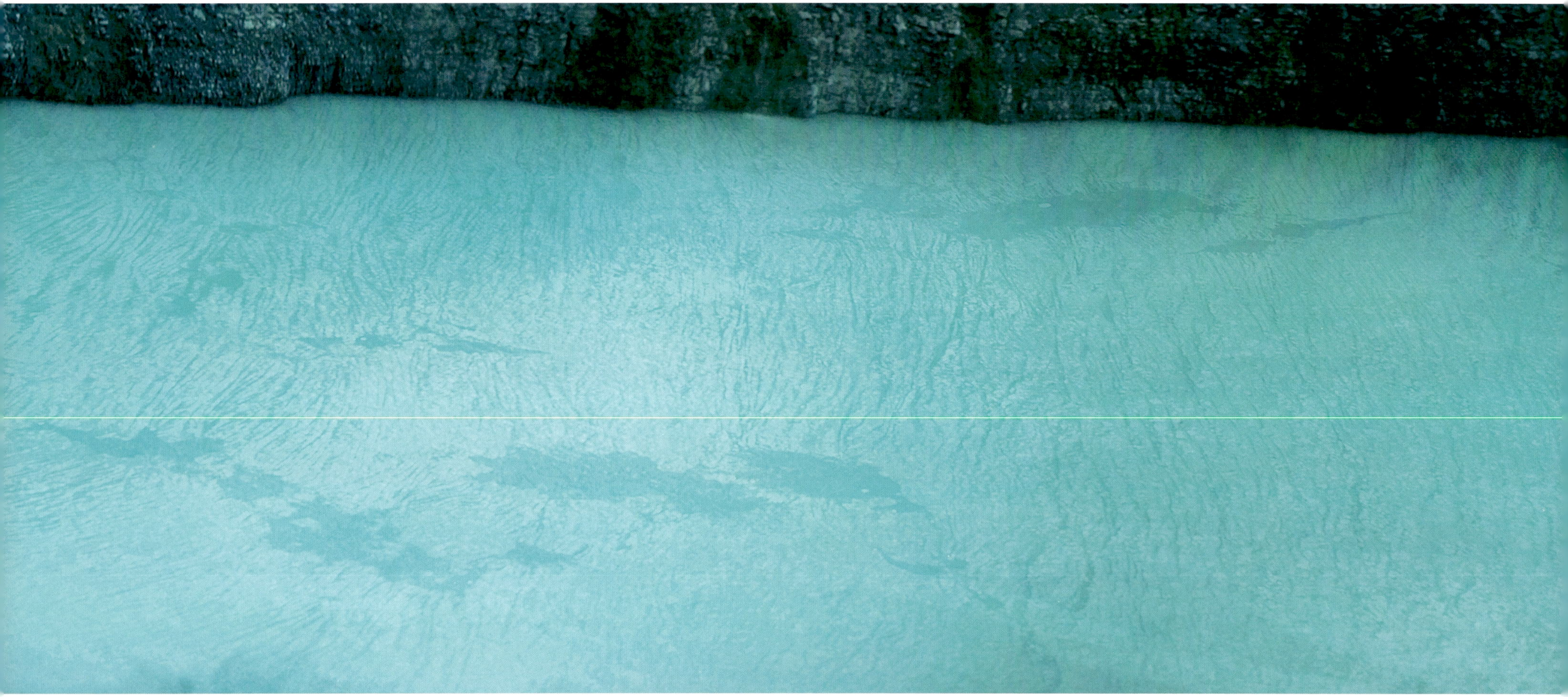

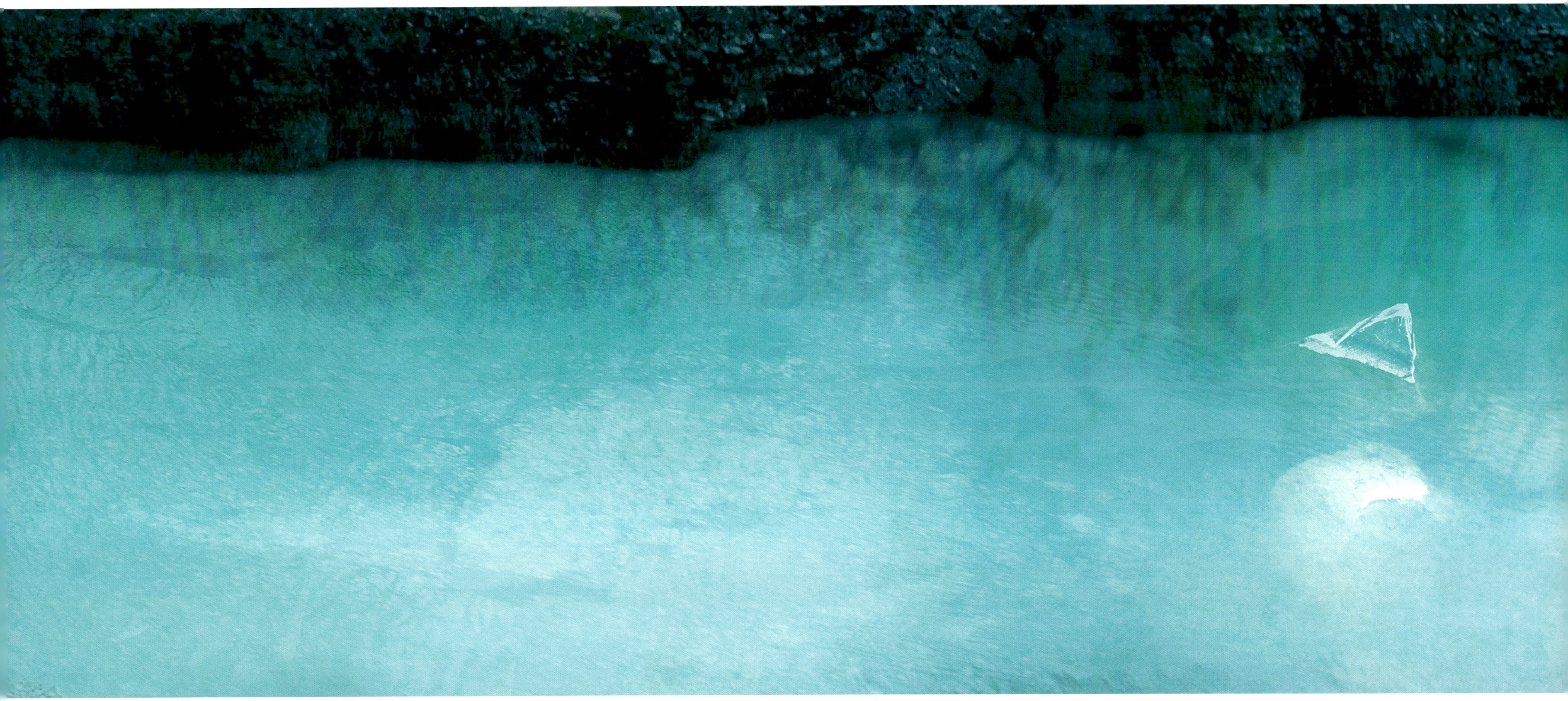

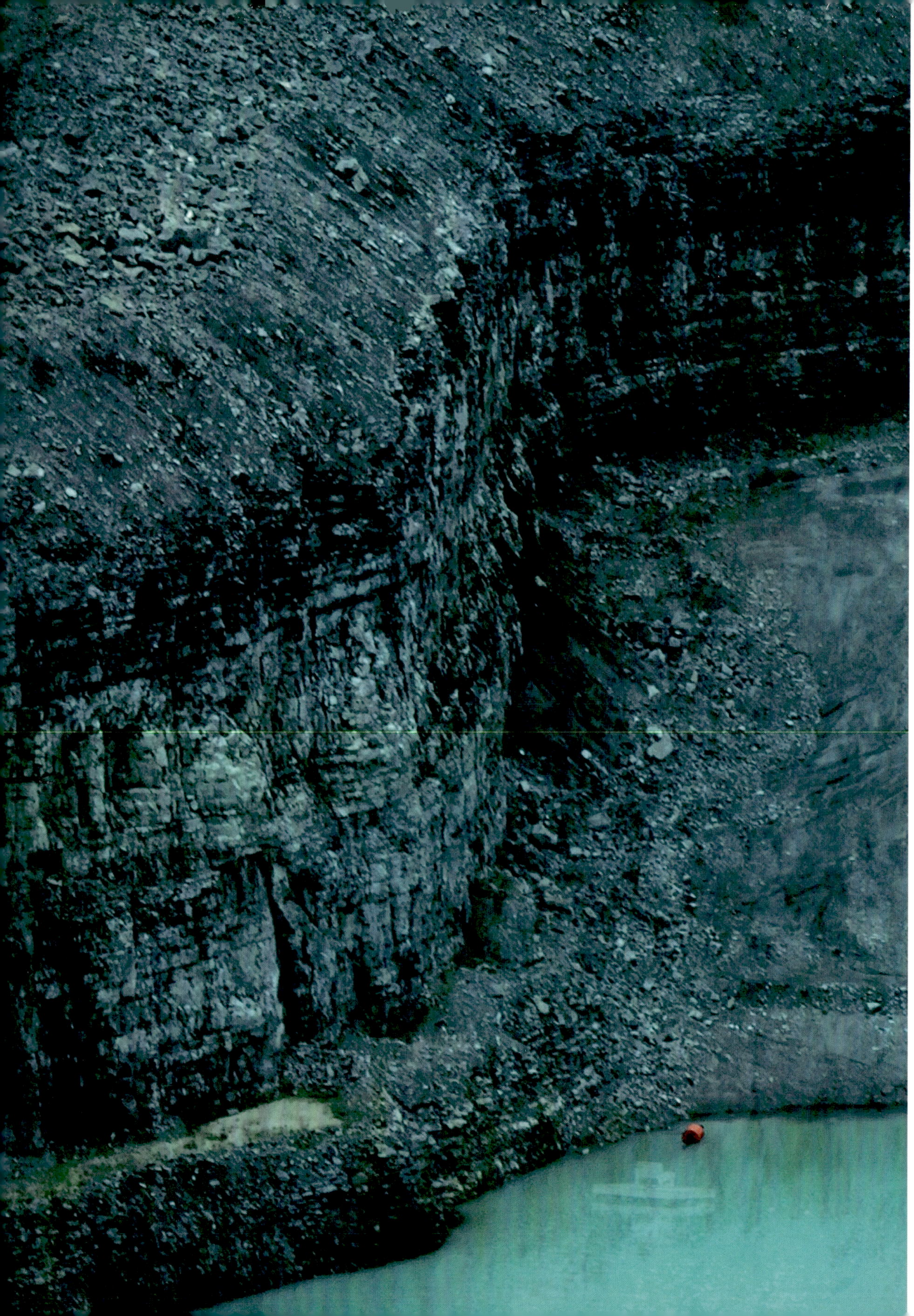

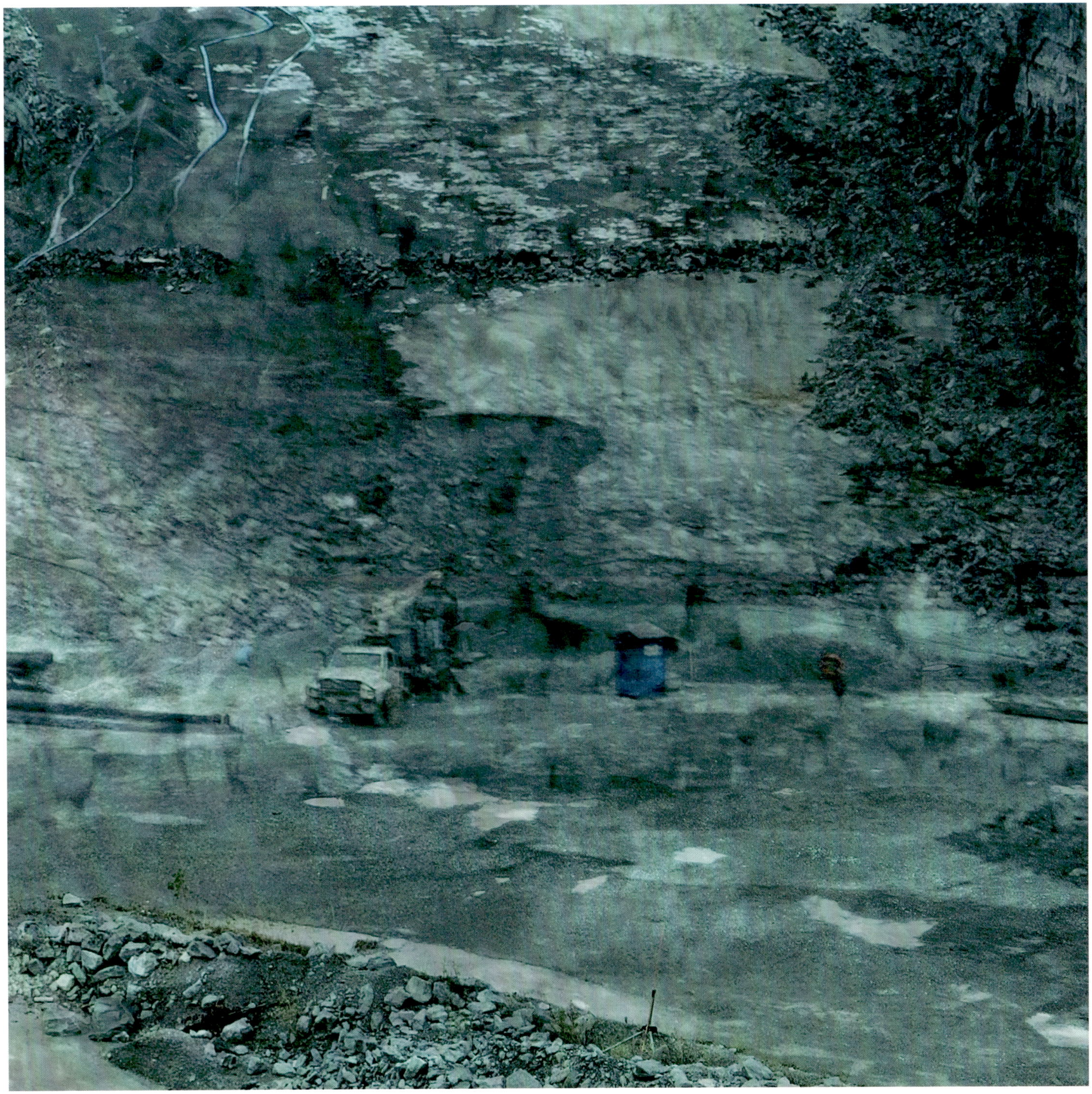

Empire Falling 14, 2012, 33.3 x 50 inches, Details pp. 50 - 53

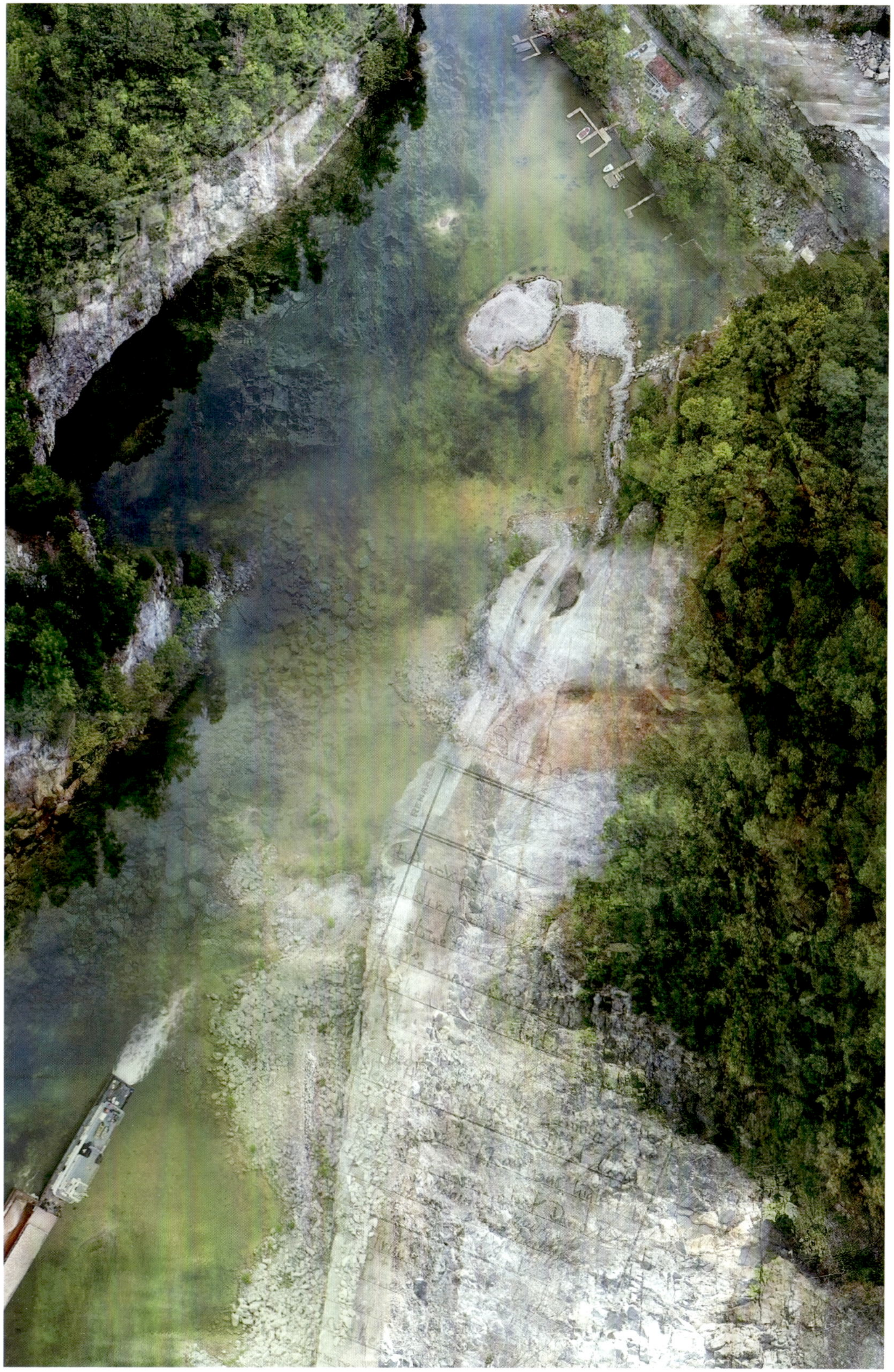

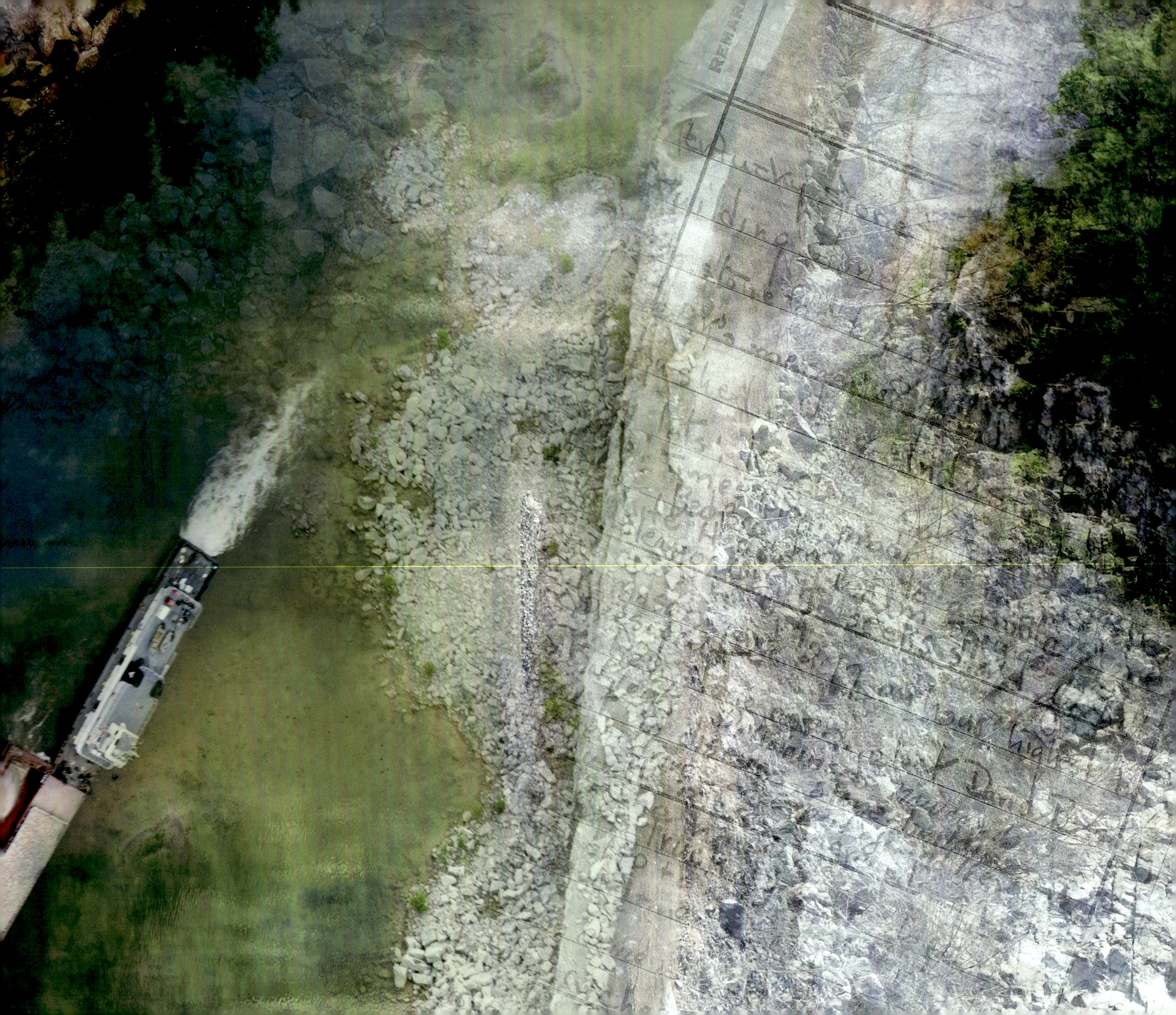

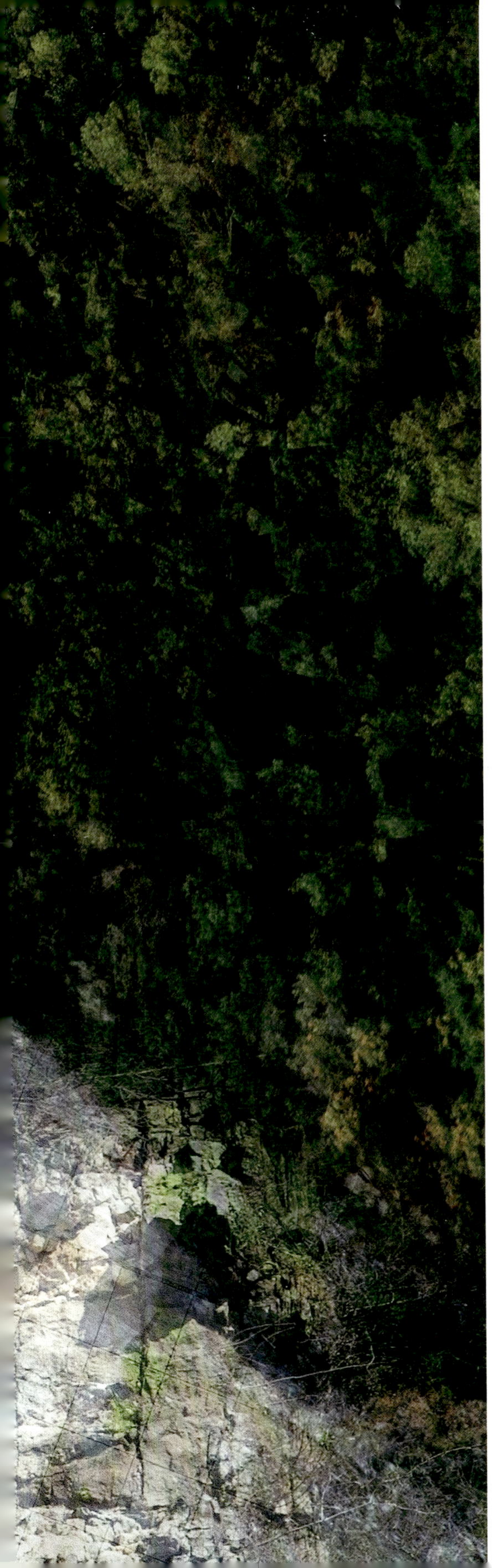

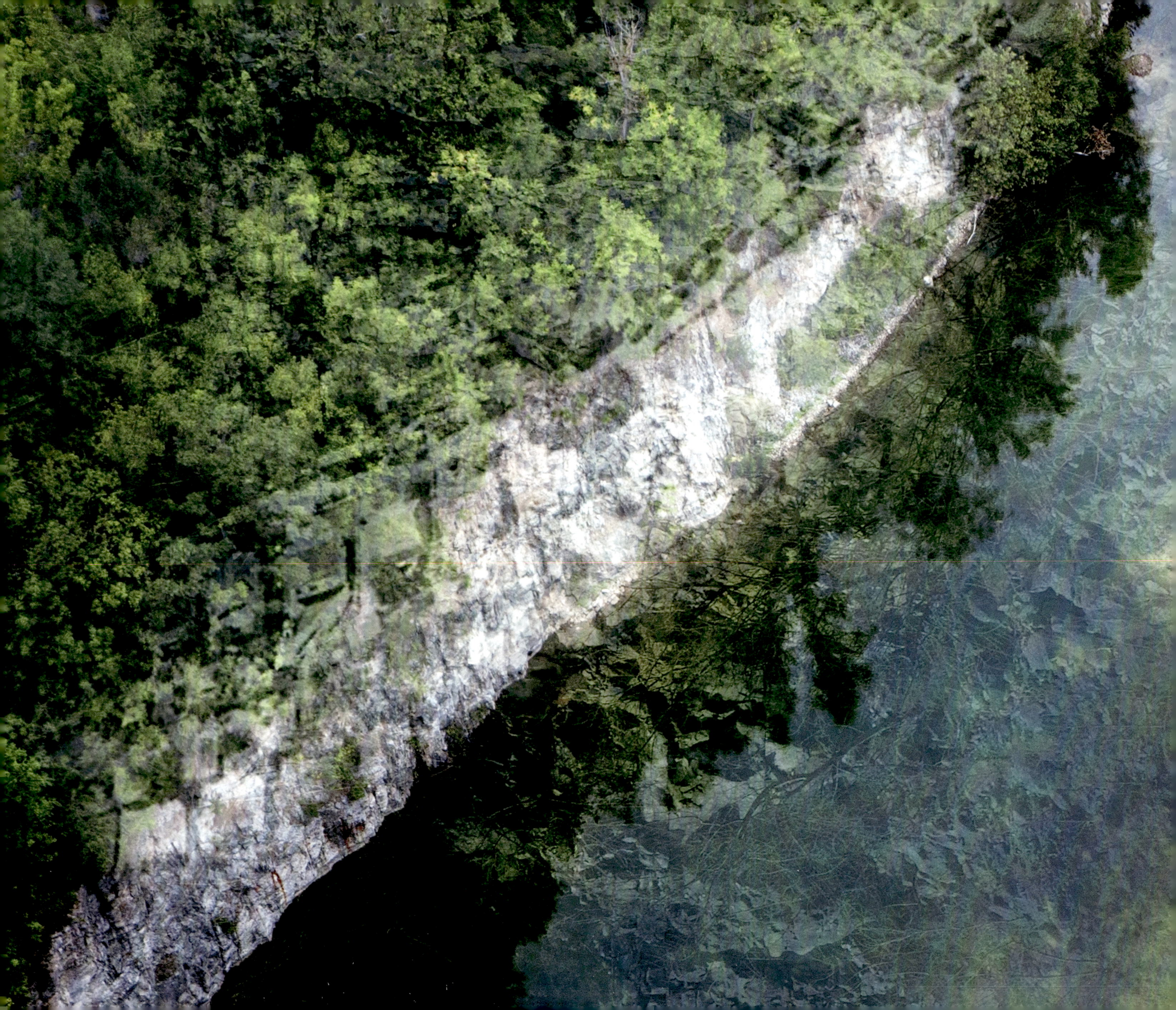

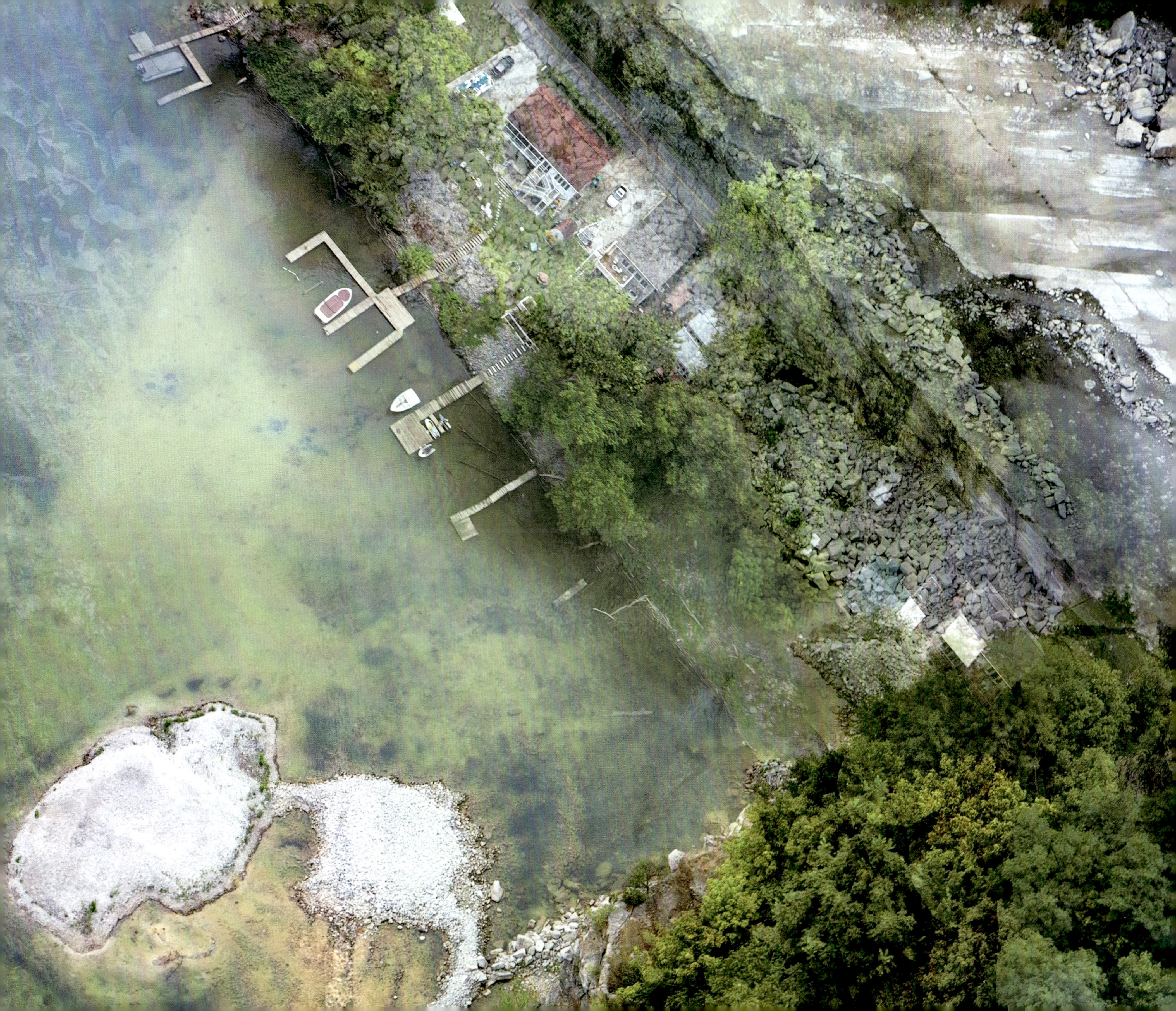

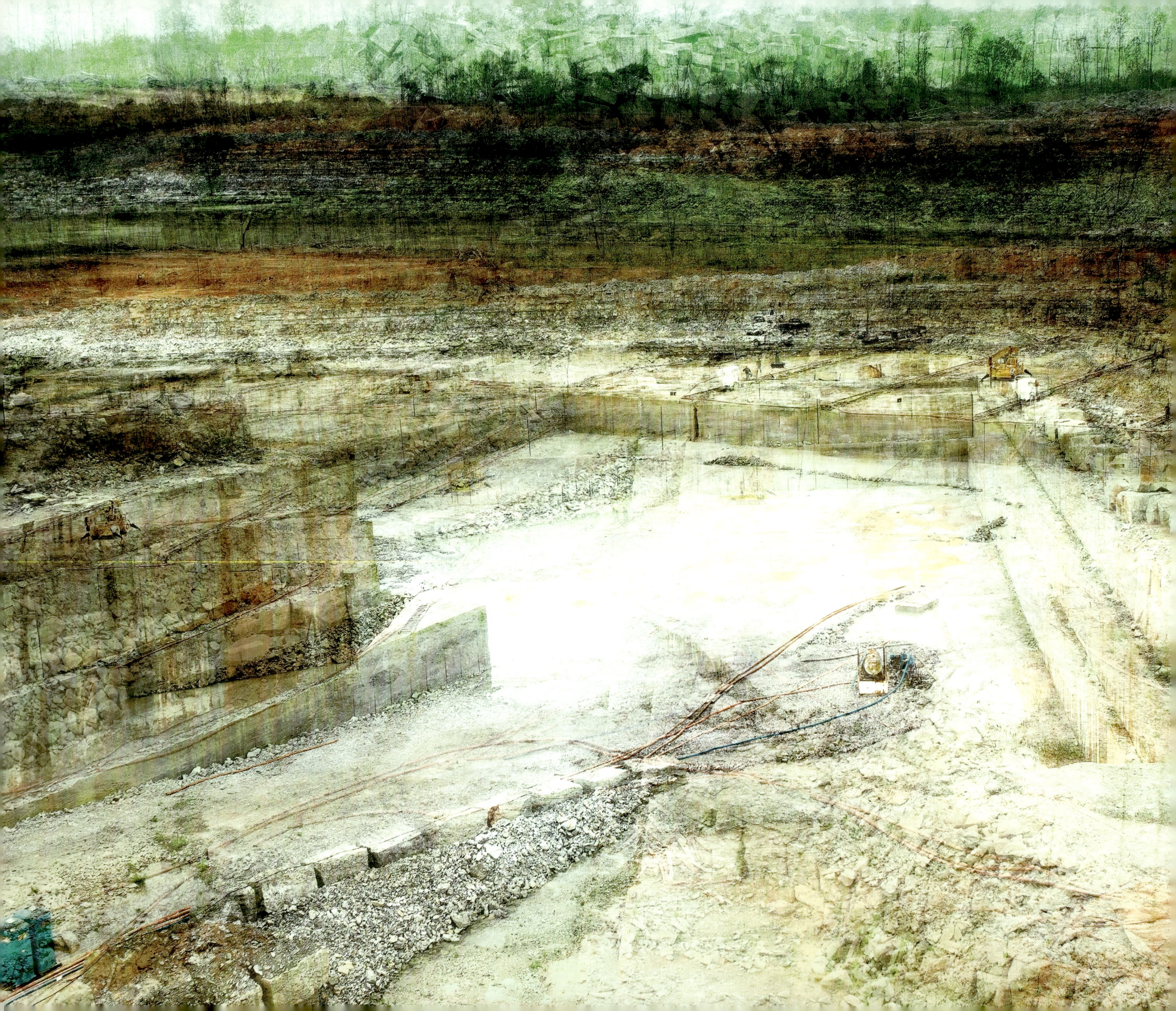

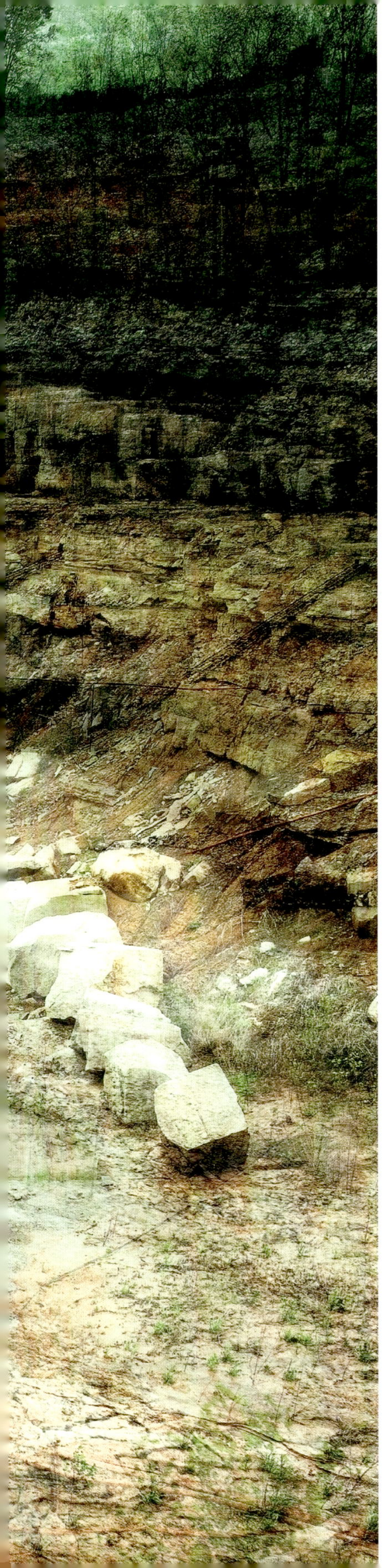

Empire Falling 1, 2012, 50 x 33 inches

Empire Falling 15, 2011, 33.3 x 50 inches, Details pp.58 - 59

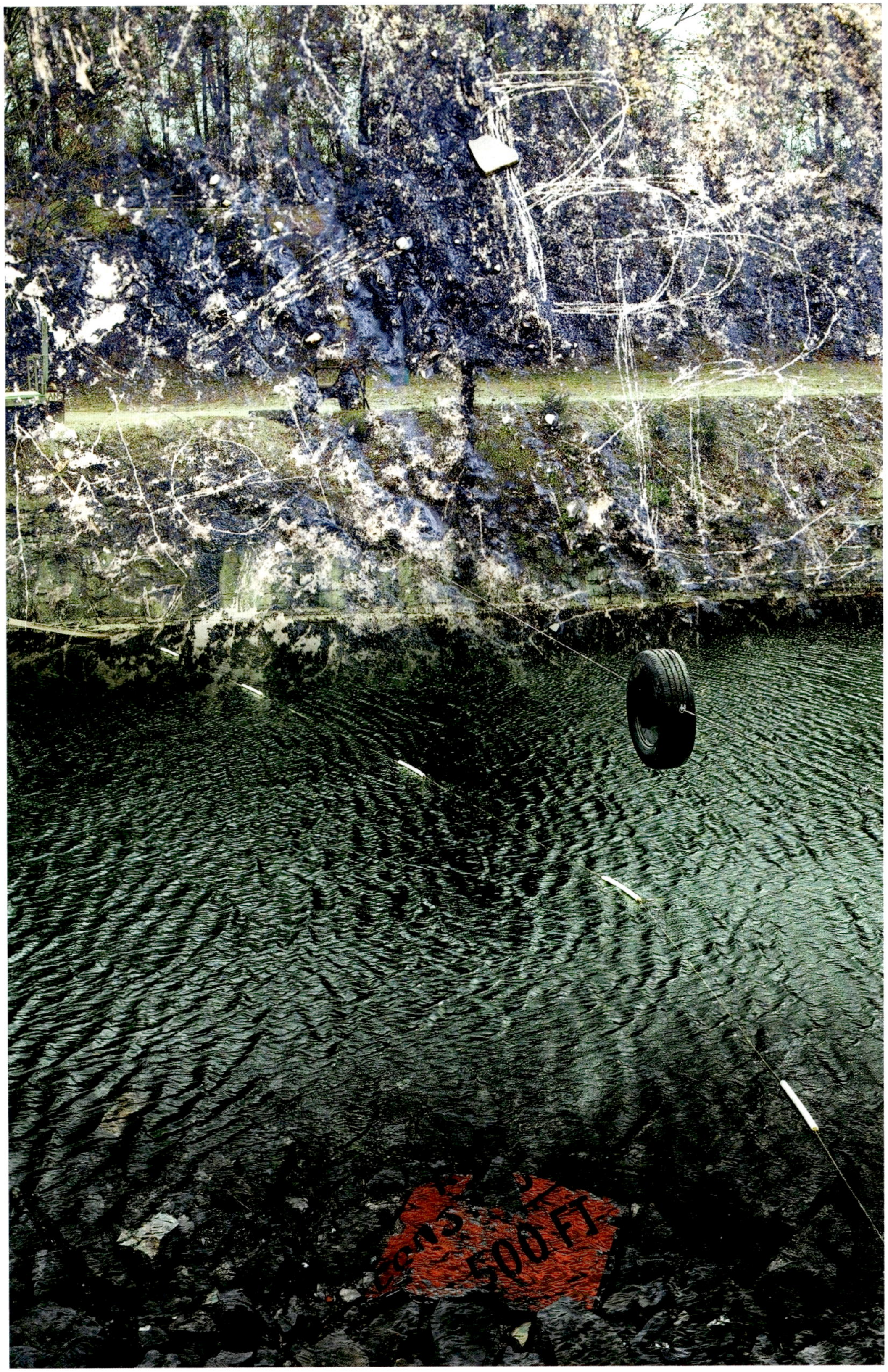
500 FT

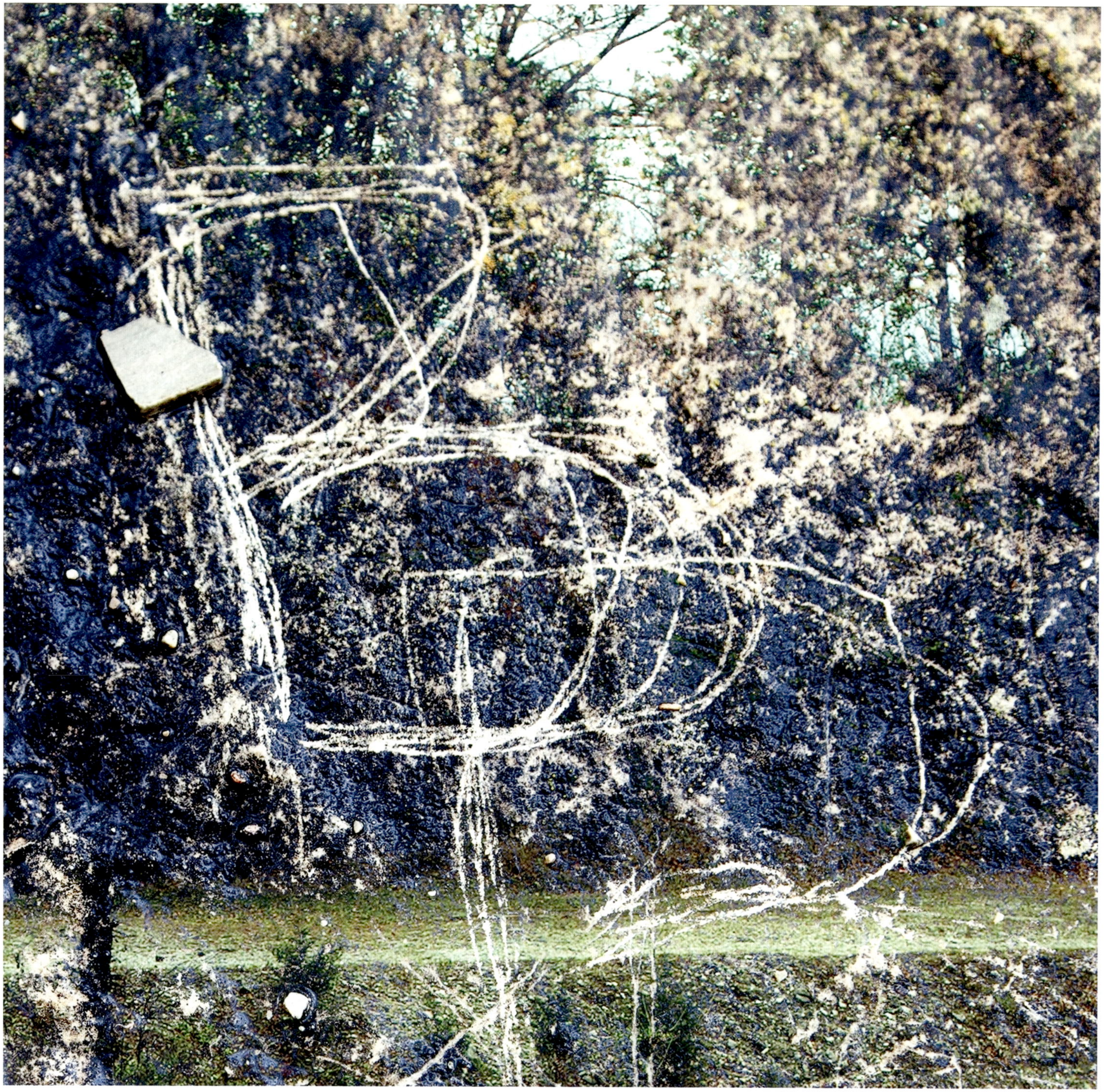

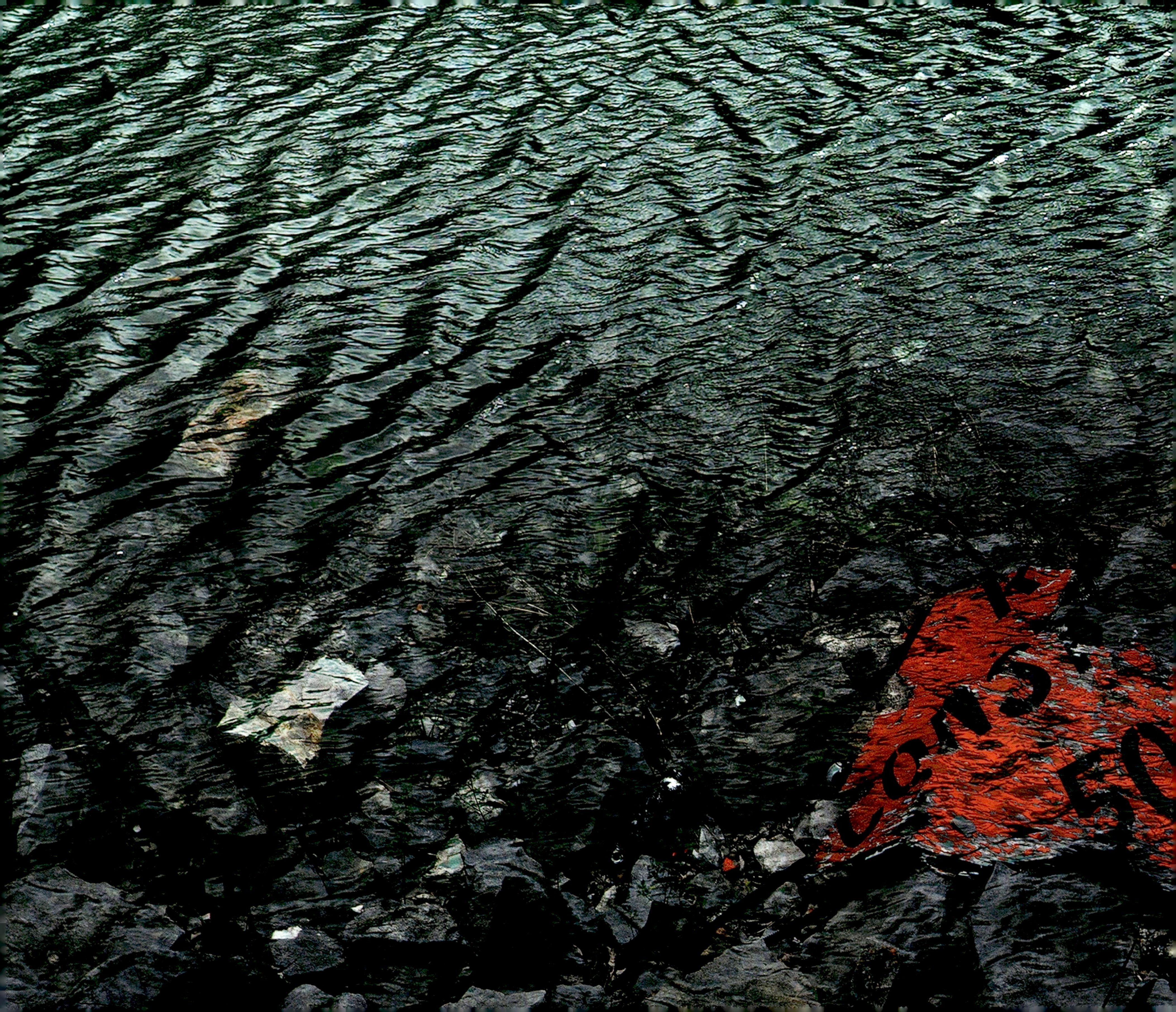

Empire Falling 16, 2012, 37.5 x 50 inches, Detail pp. 62 - 63

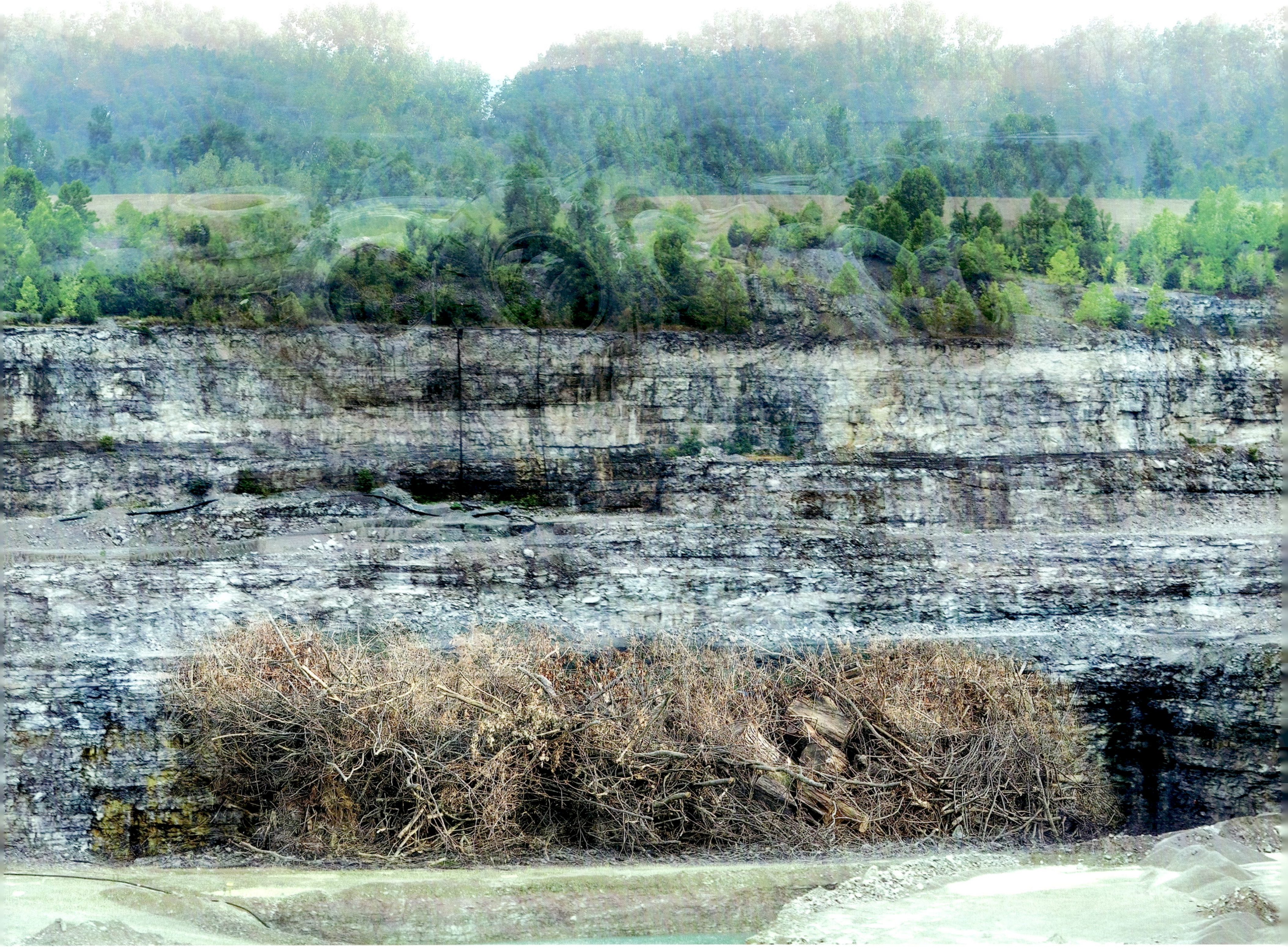

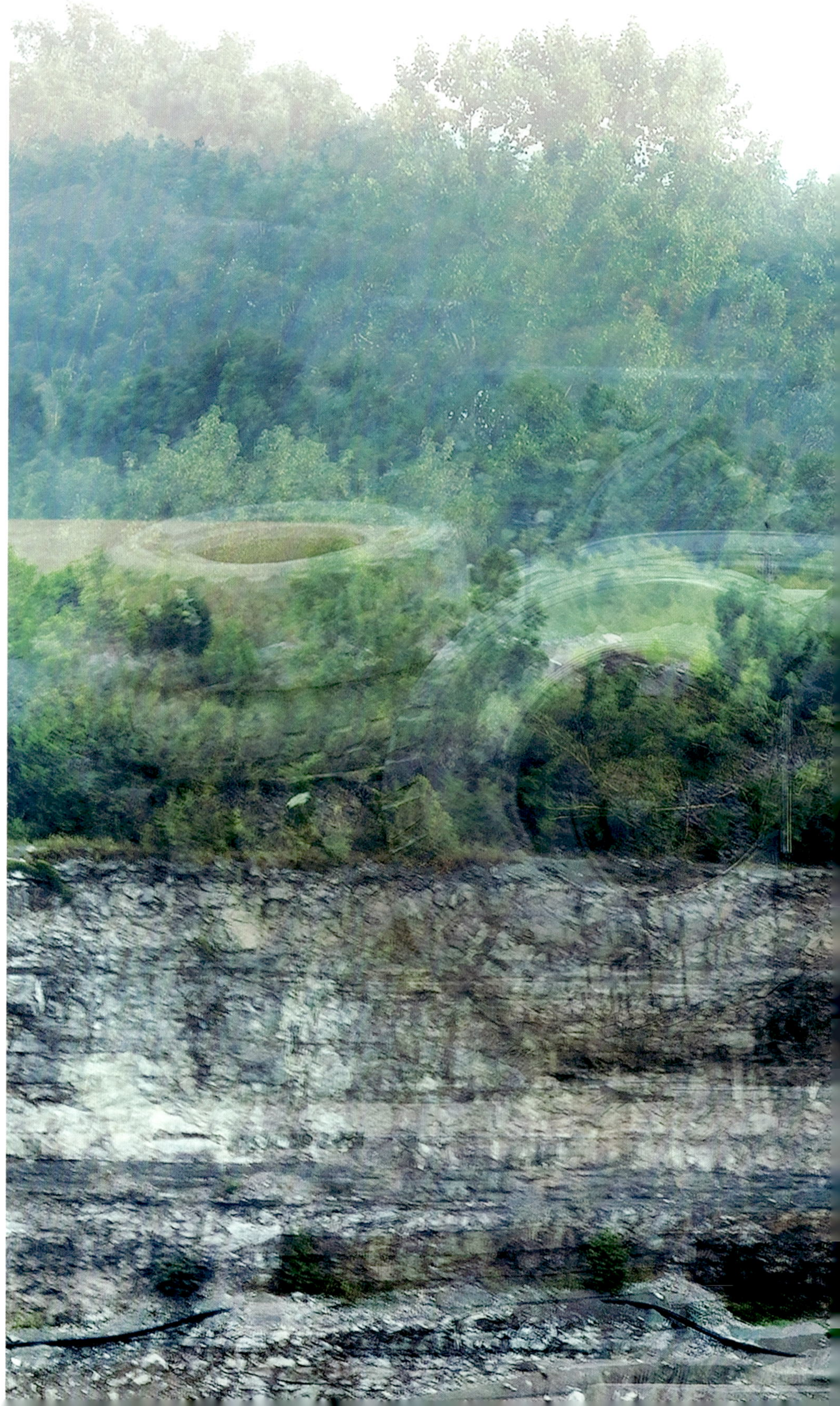

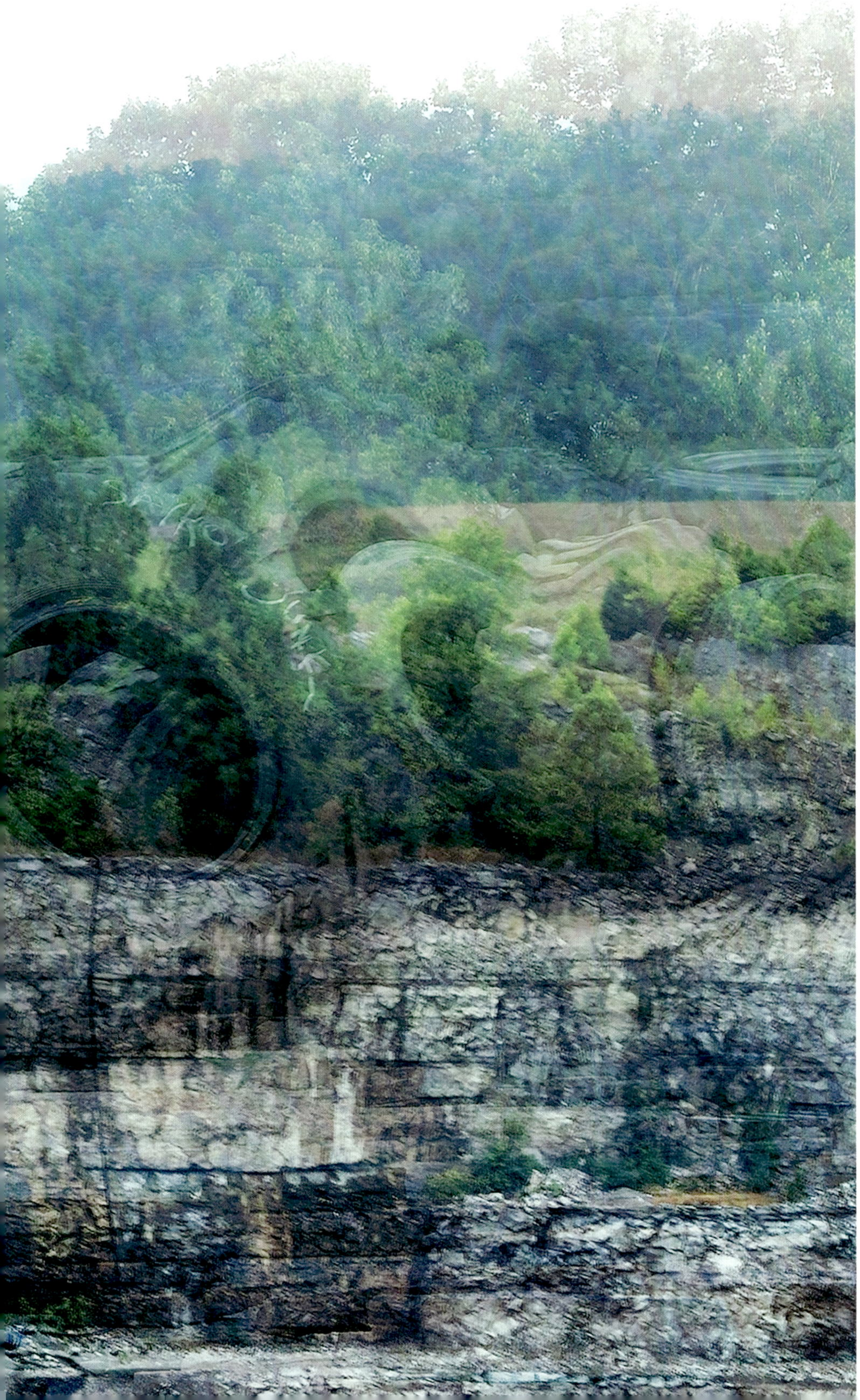

Empire Falling 17, 2012, 37.5 x 50 inches, Detail pp. 66 - 67

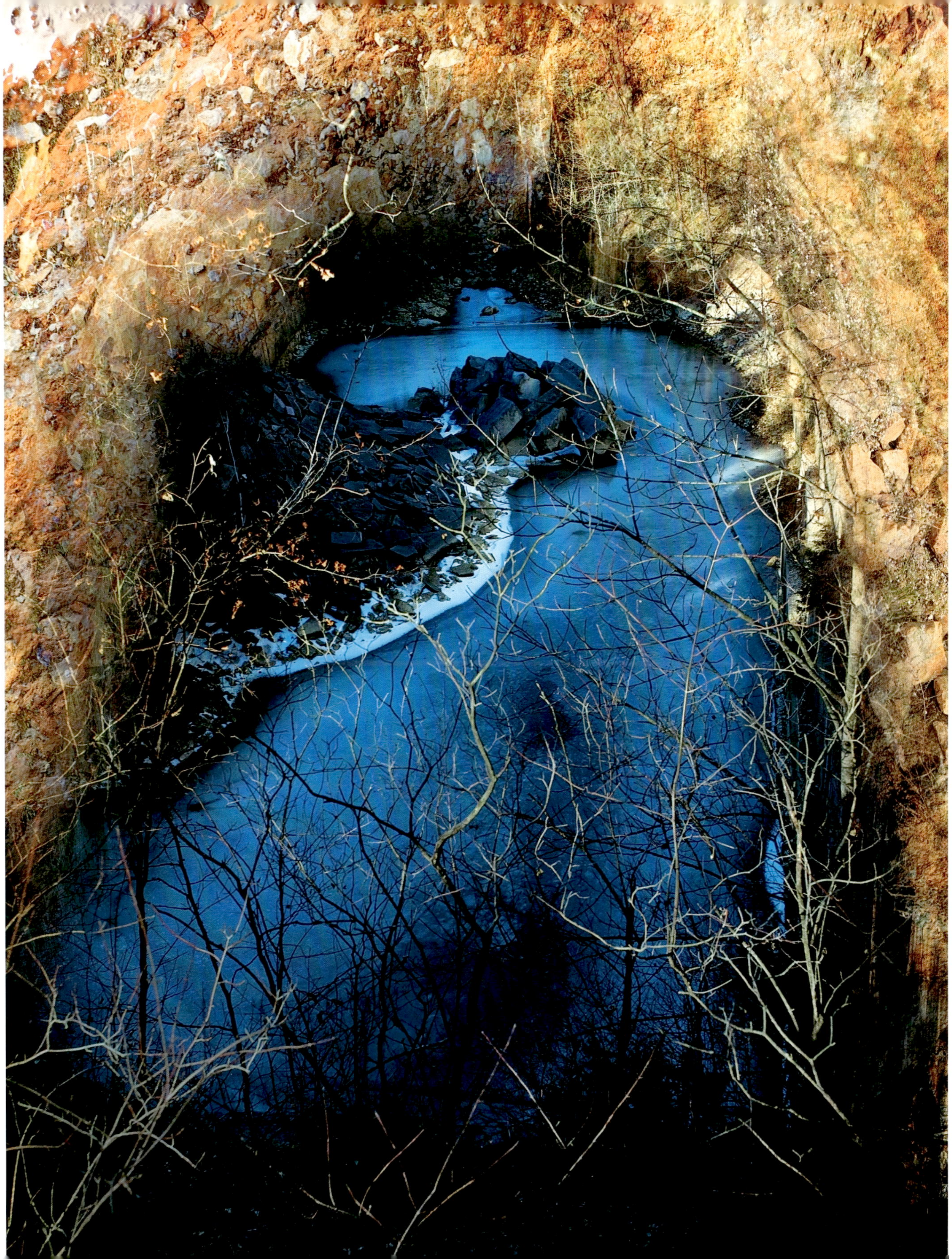

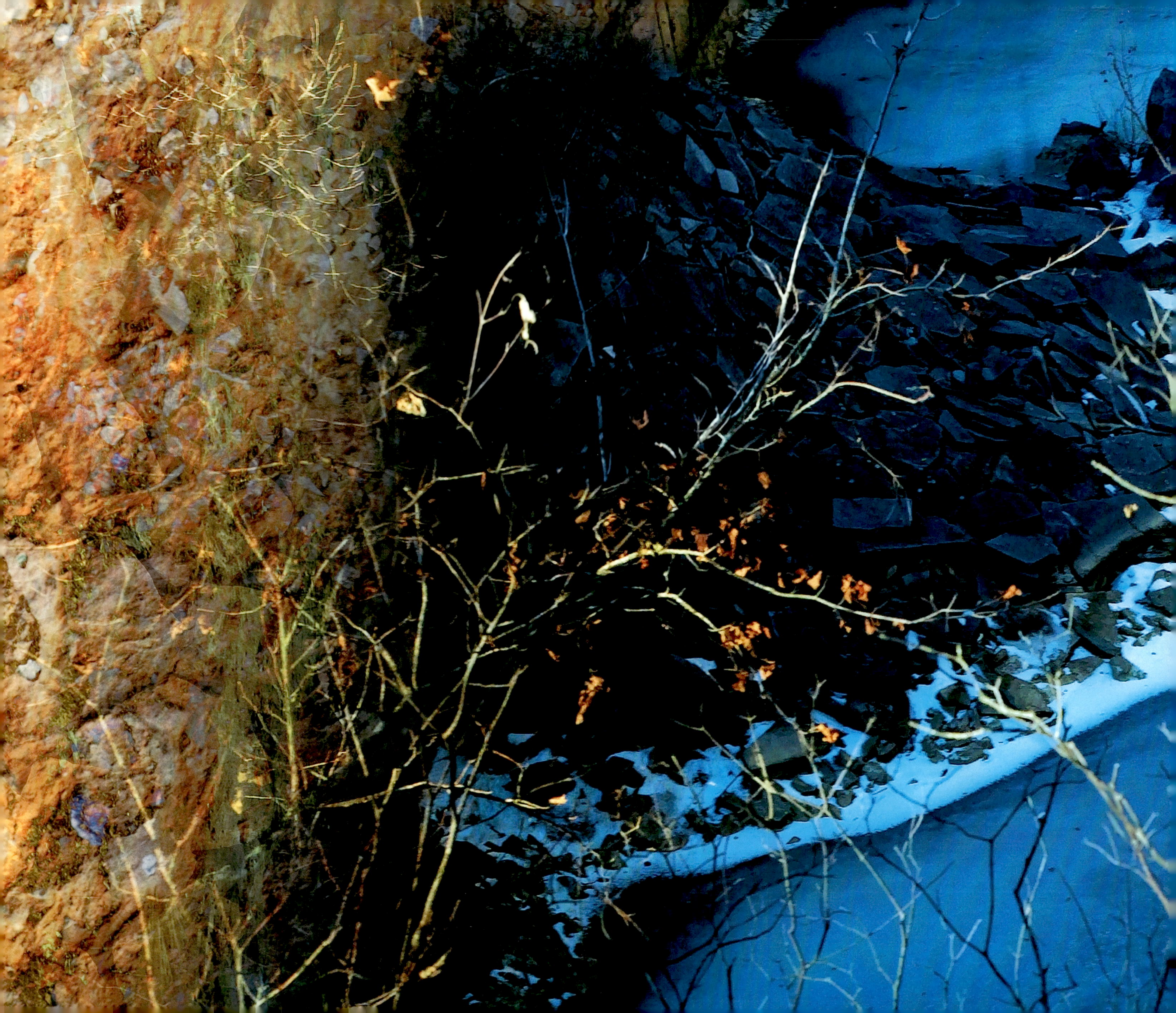

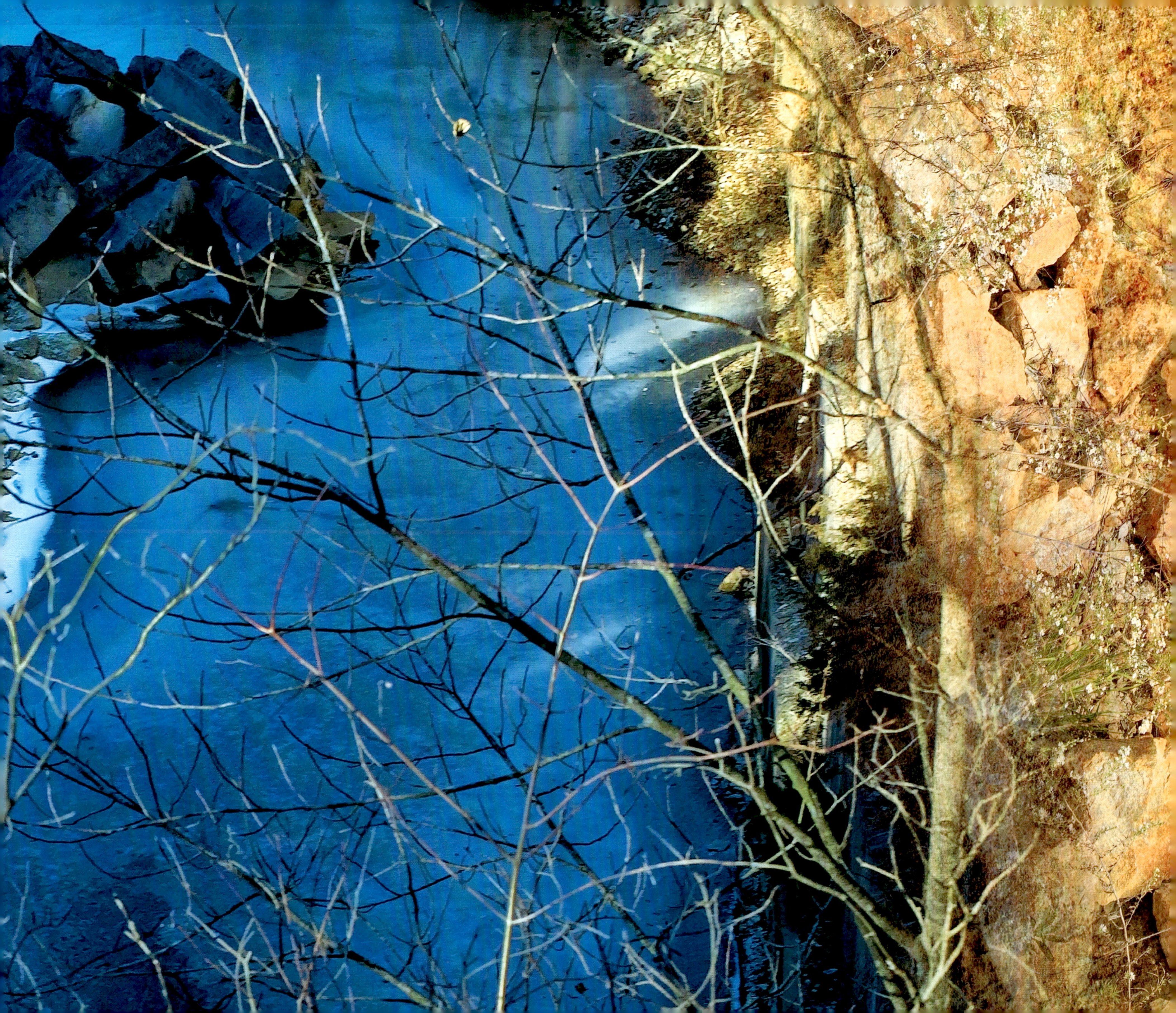

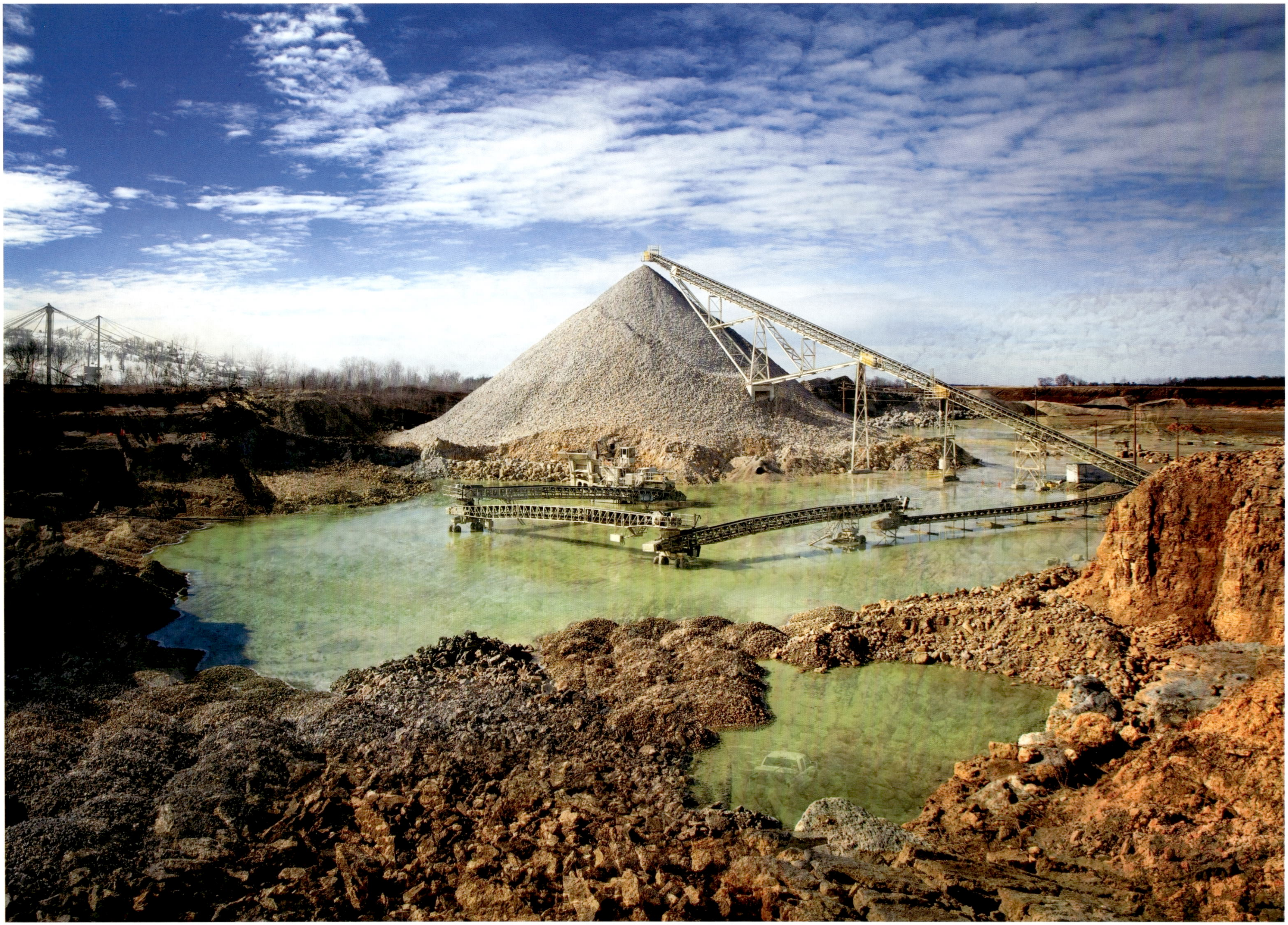

Empire Falling 18, 2012, 33.3 x 50 inches, Details p. 69

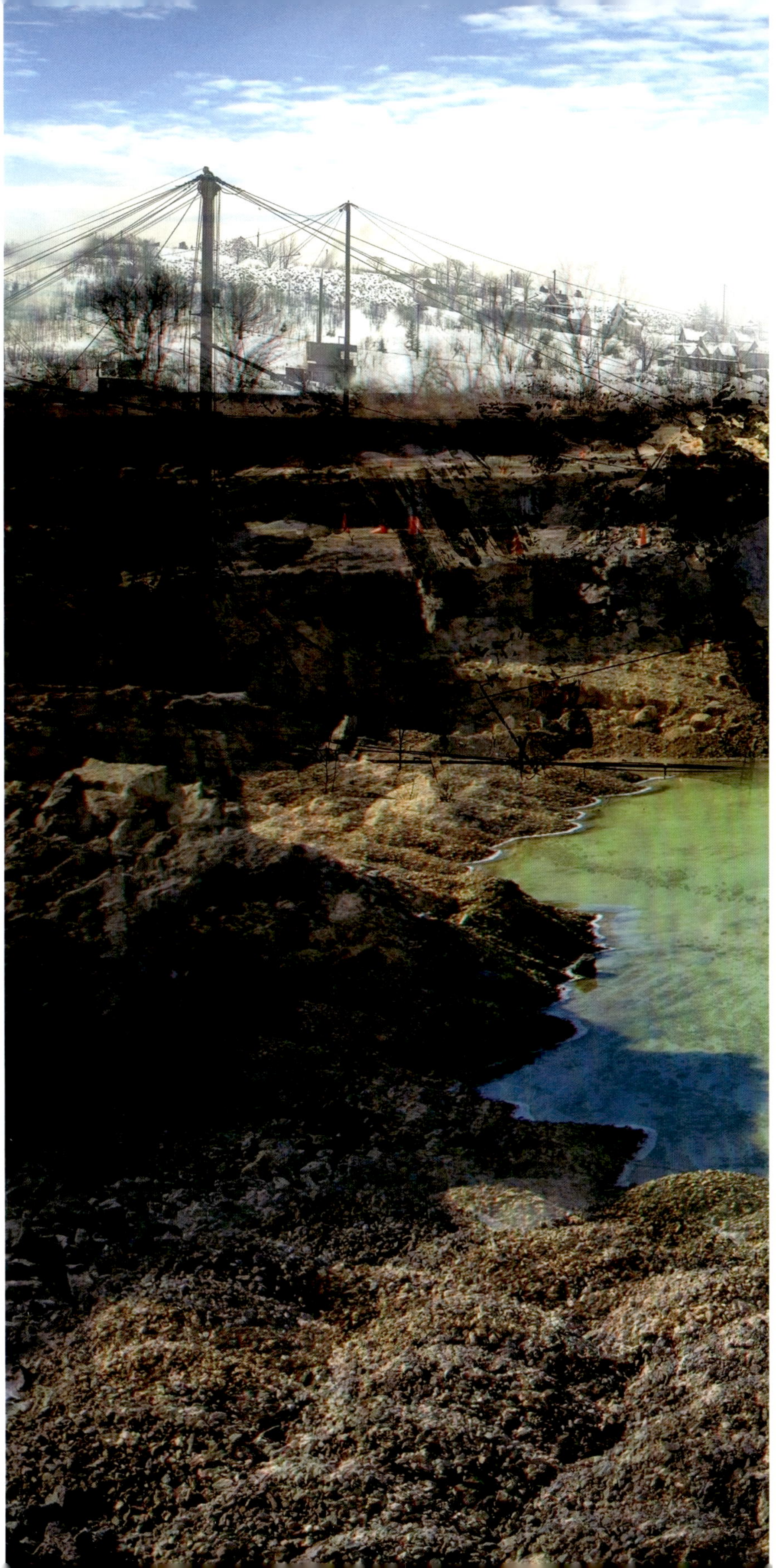

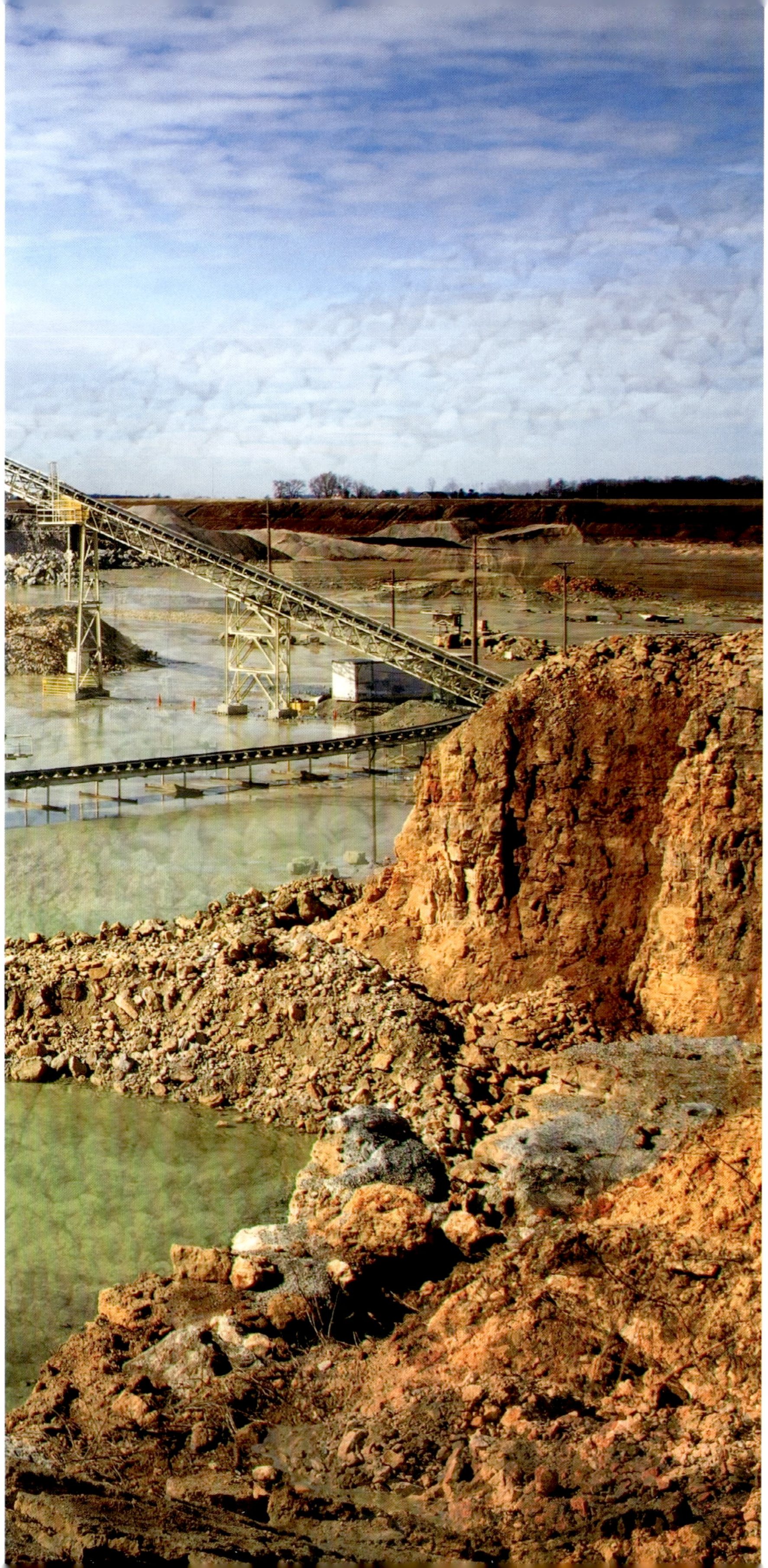

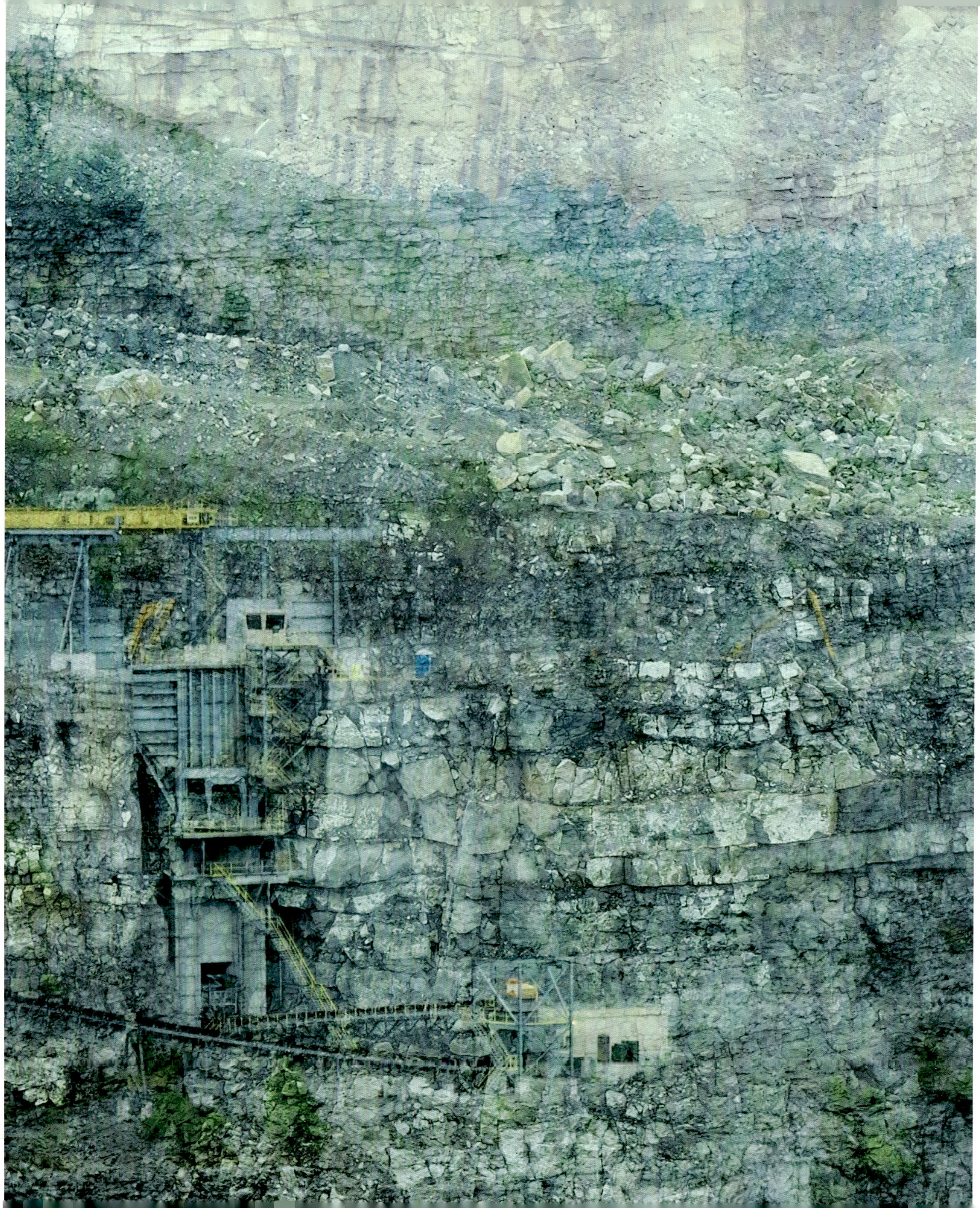

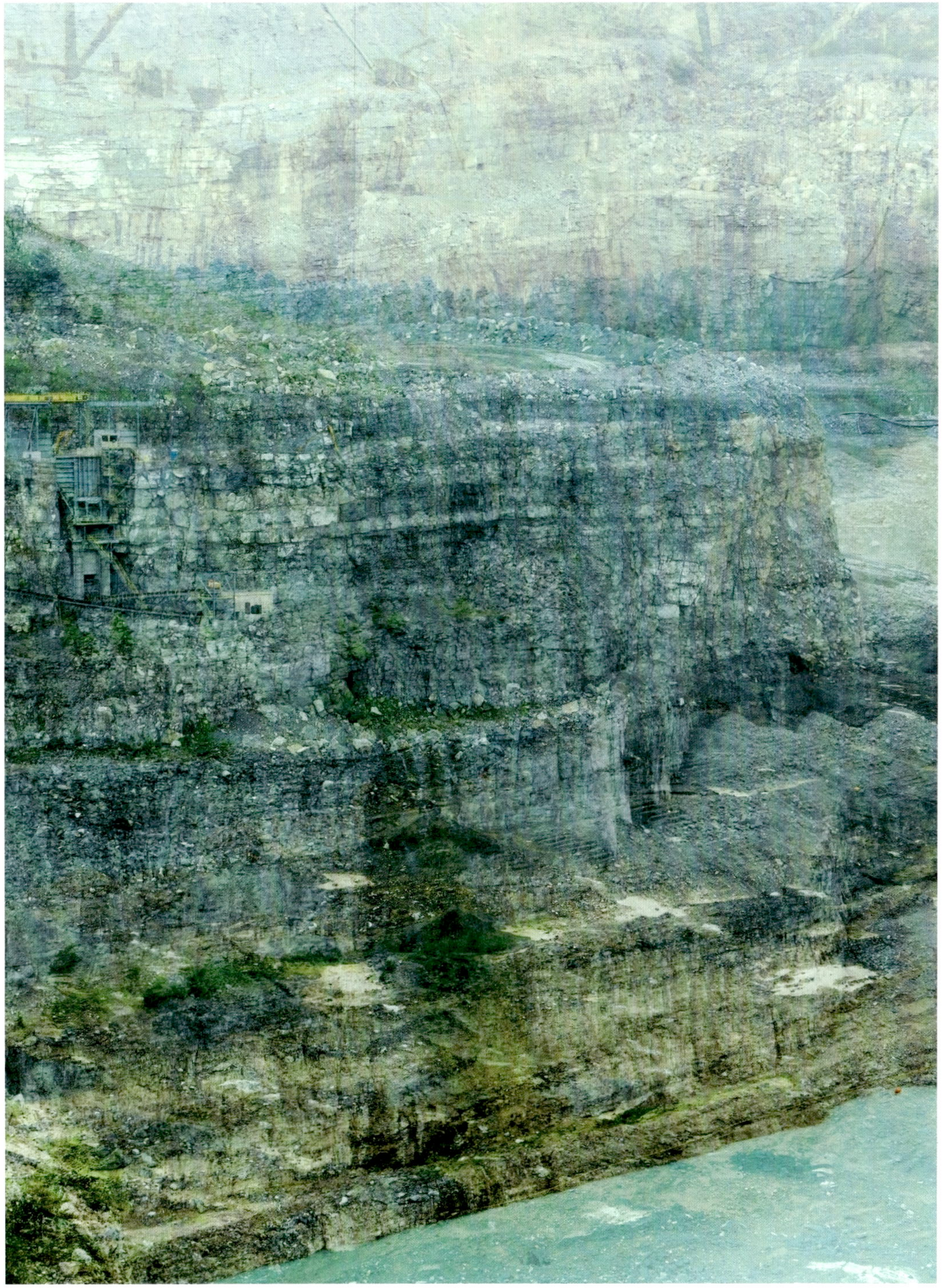

Empire Falling 19, 2012, 33.3 x 50 inches, Detail p. 70

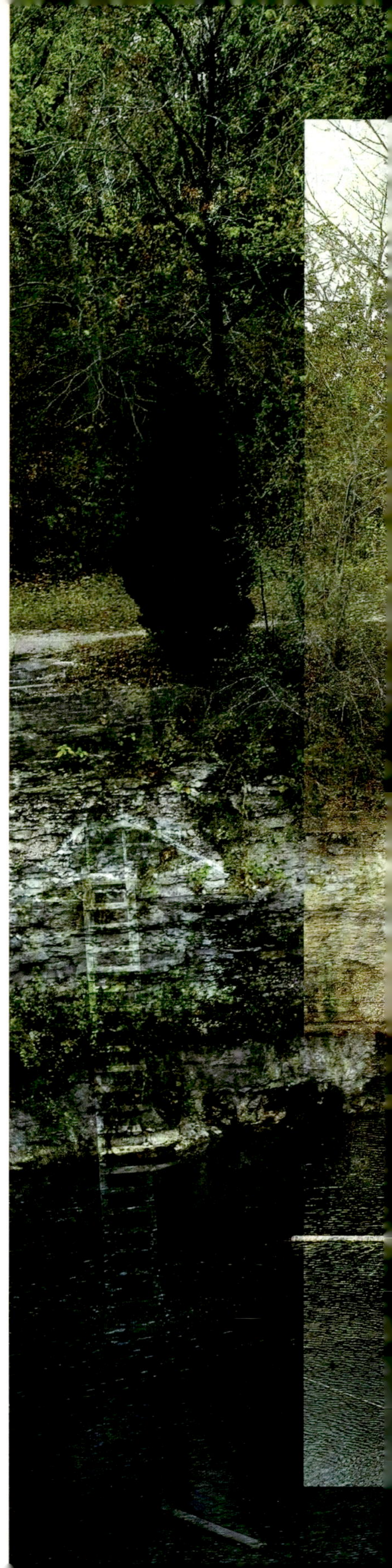

Empire Falling 20, 2011, 33.3 x 50 inches, Details pp. 74 - 75

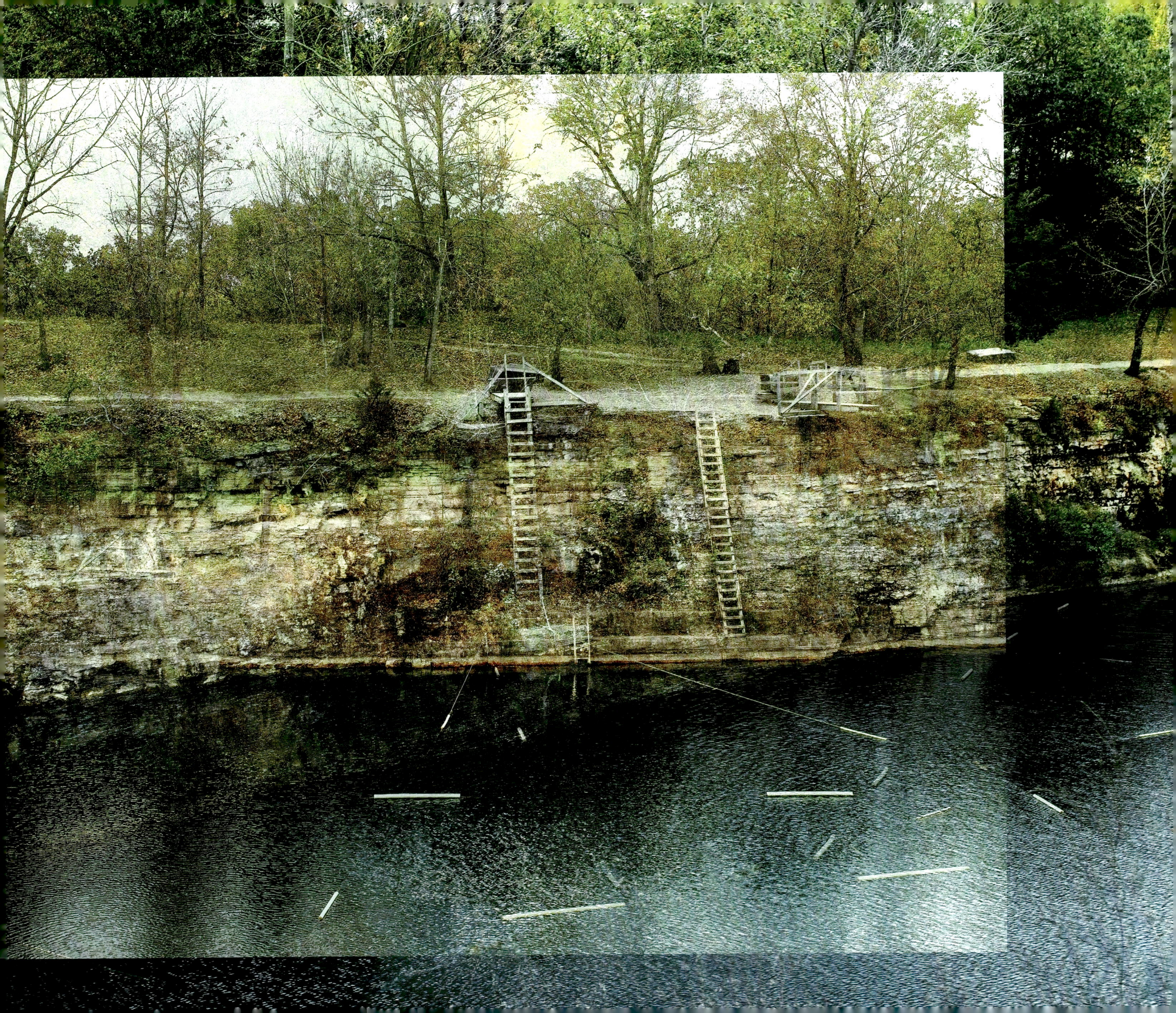

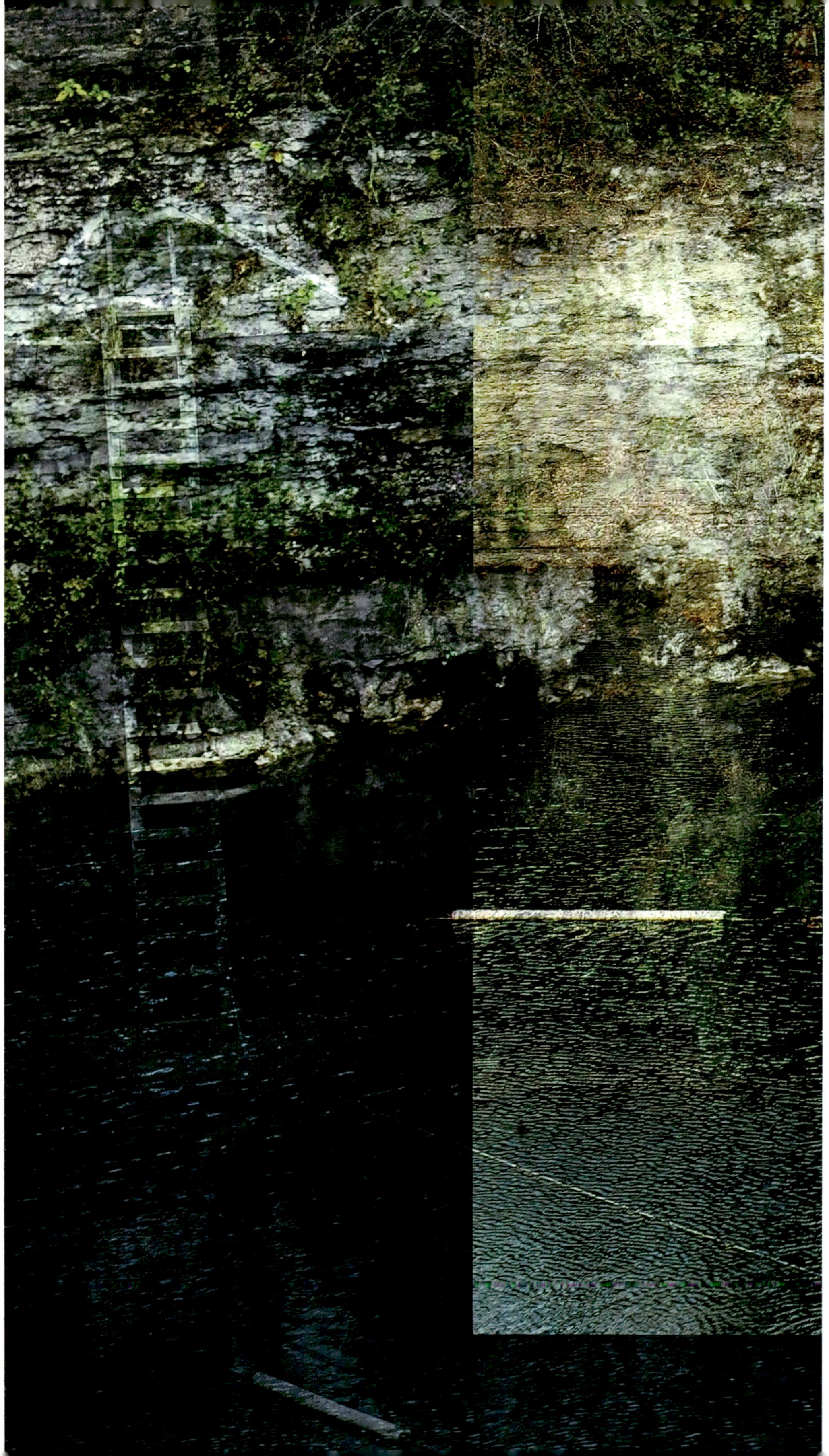

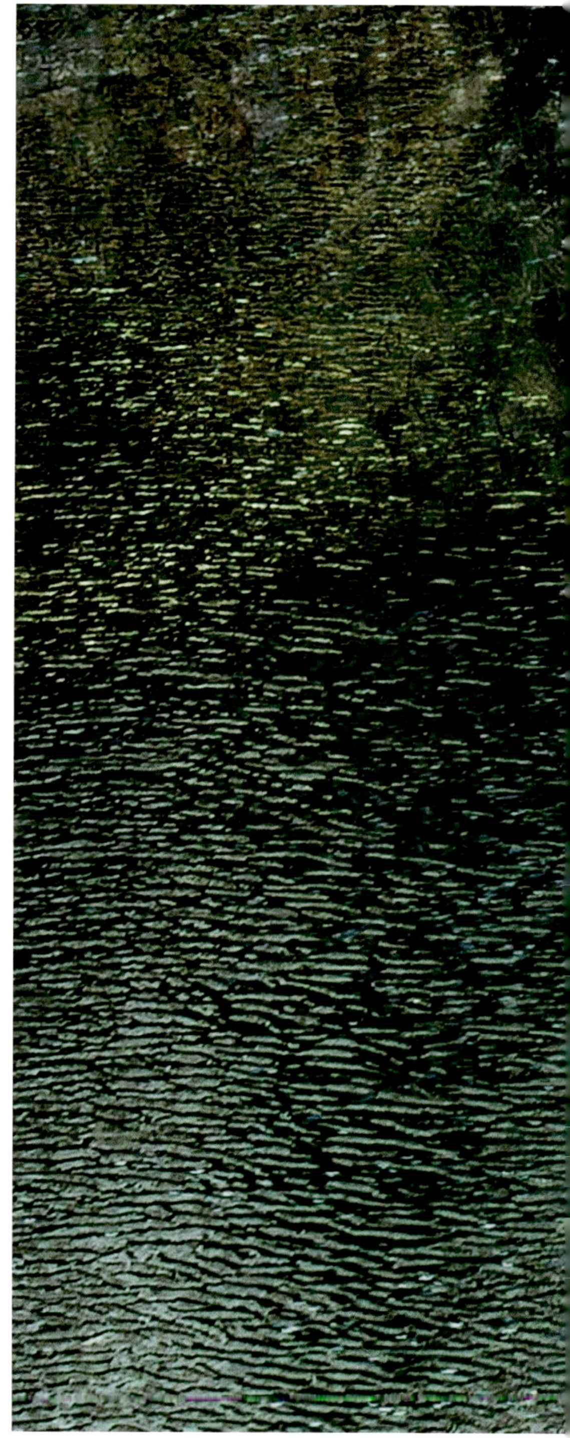

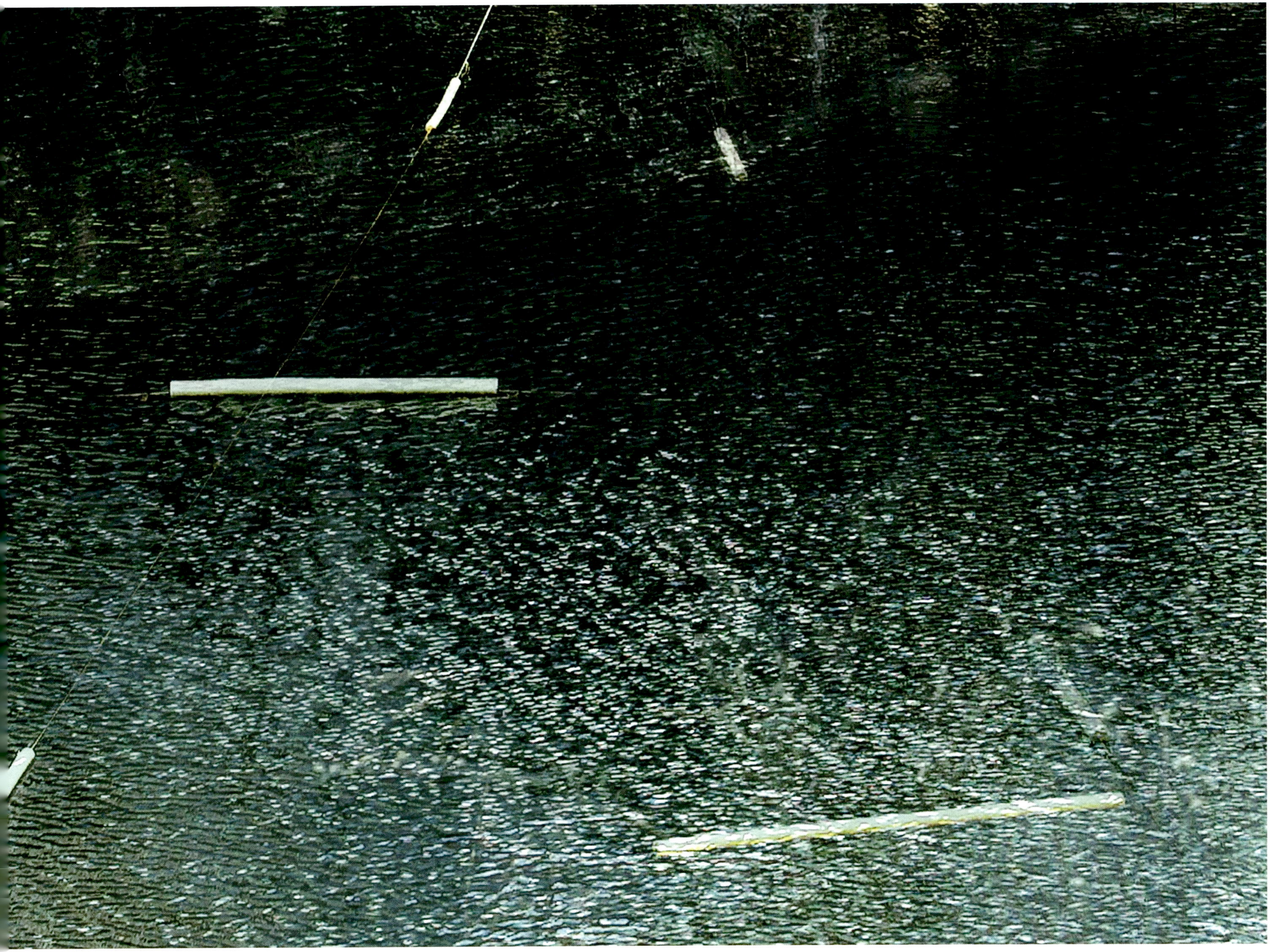

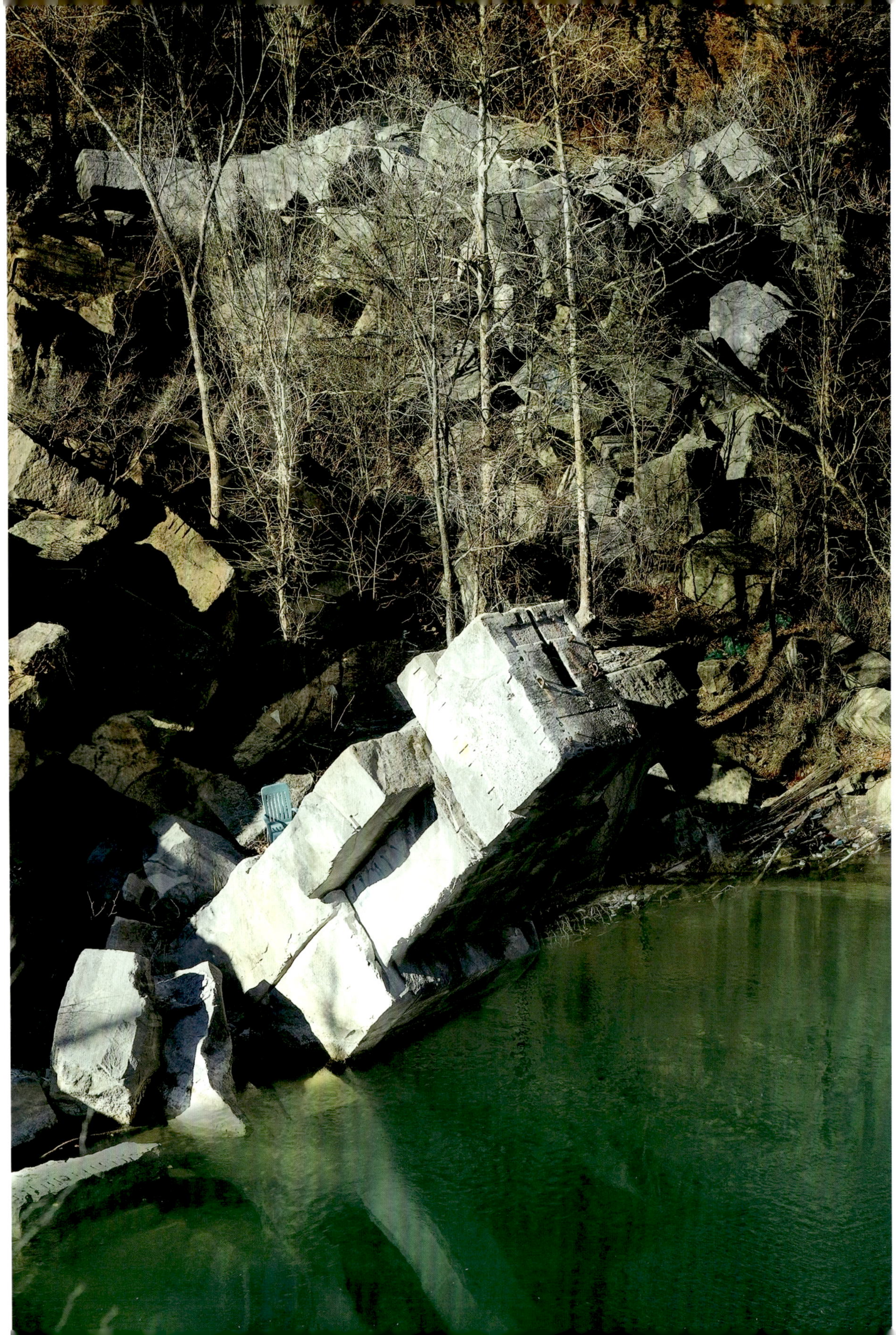

Empire Falling 22, 2012, 33.3 x 50 inches

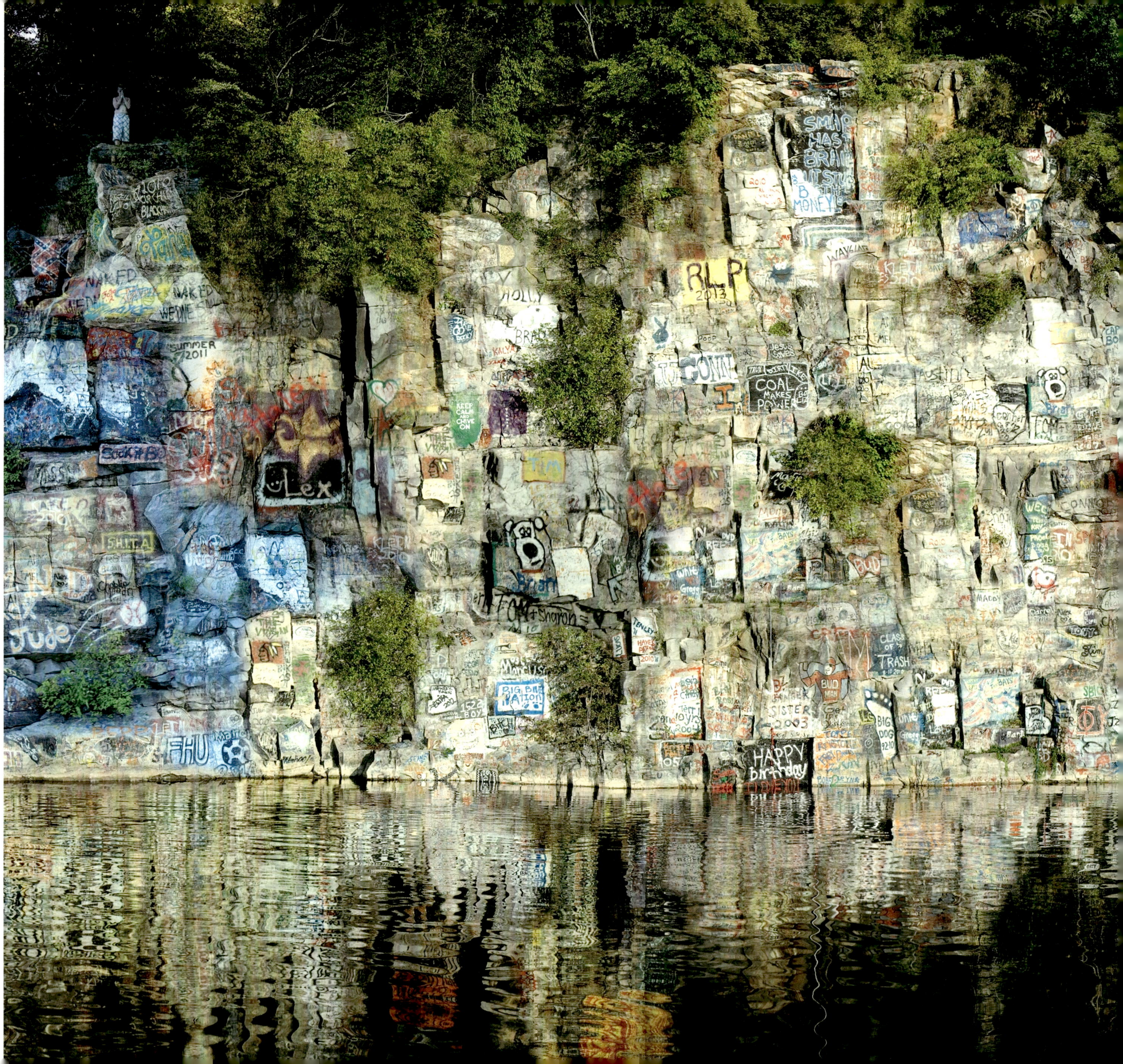
RLP
2013
COAL
MAKES
POWER
HAPPY
birthday
SISTER
2003
Lex
SUMMER
2011
HOLLY
Jude
FHU
CLASS
OF
TRASH
BIG
DOG
MONEY

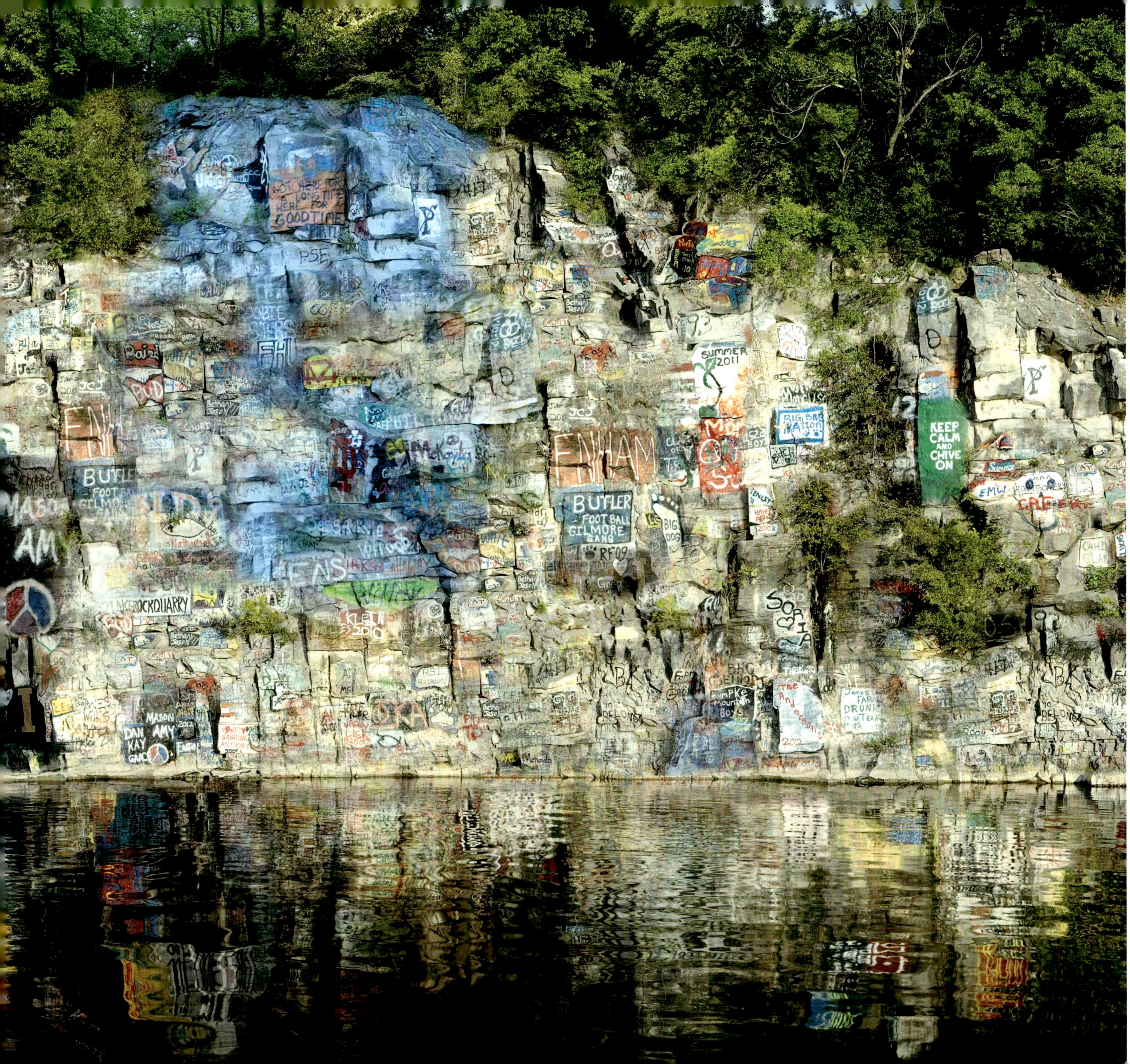
KEEP CALM AND CHIVE ON
BUTLER FOOTBALL GILMORE GANG
SUMMER 2011
KLEIN 2010

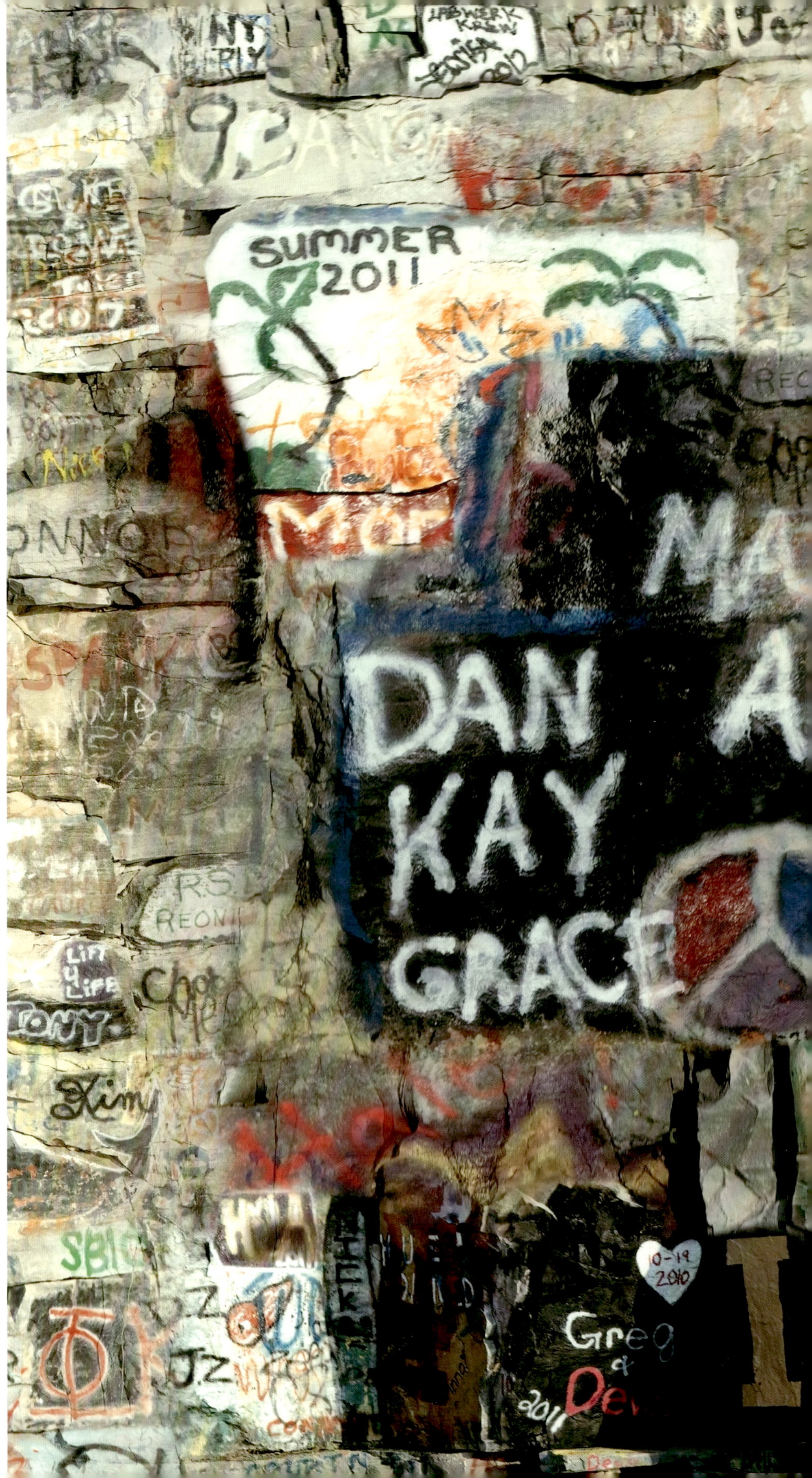
SUMMER
2011
DAN
KAY
GRACE
10-19
2010
Greg

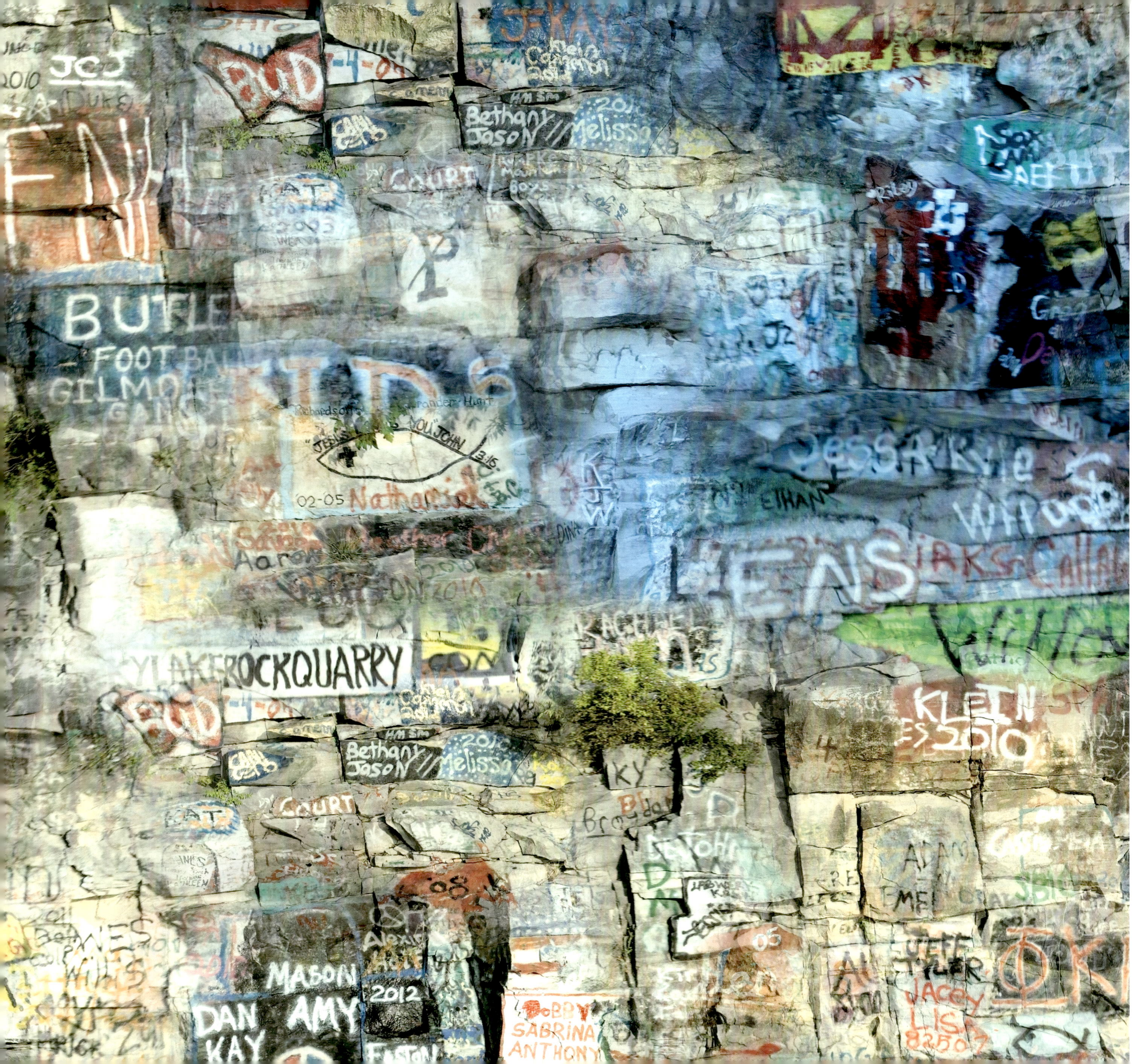
BUTLER
Bethany
Jason
Melissa
COURT
ETHAN
ROCKQUARRY
KLEIN
MASON
AMY
DAN
KAY
2012
SABRINA
ANTHONY

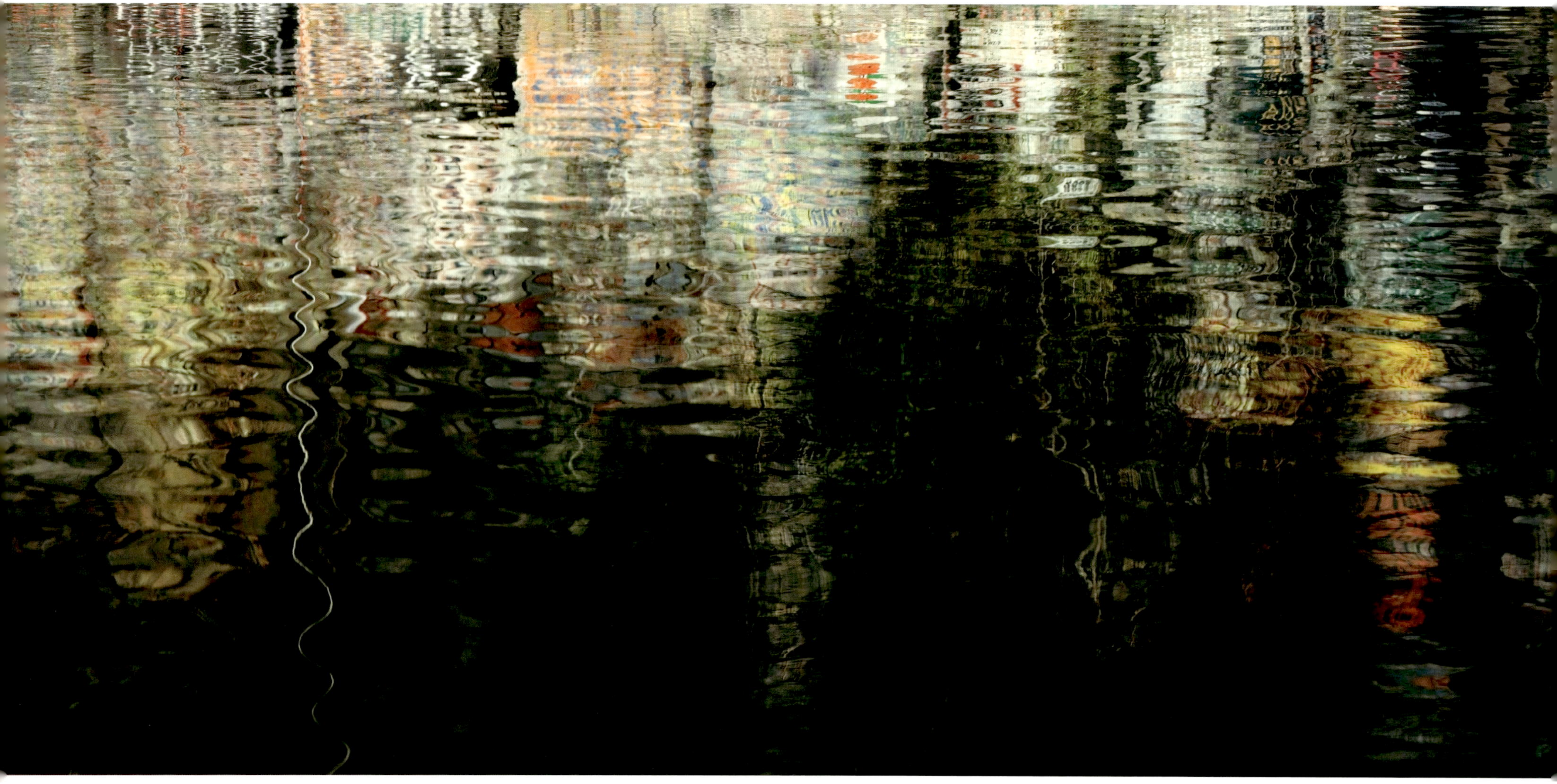

Page 78 - 79, Empire Falling 21, 2012, 48 x 107 inches, Details pp. 80 - 83

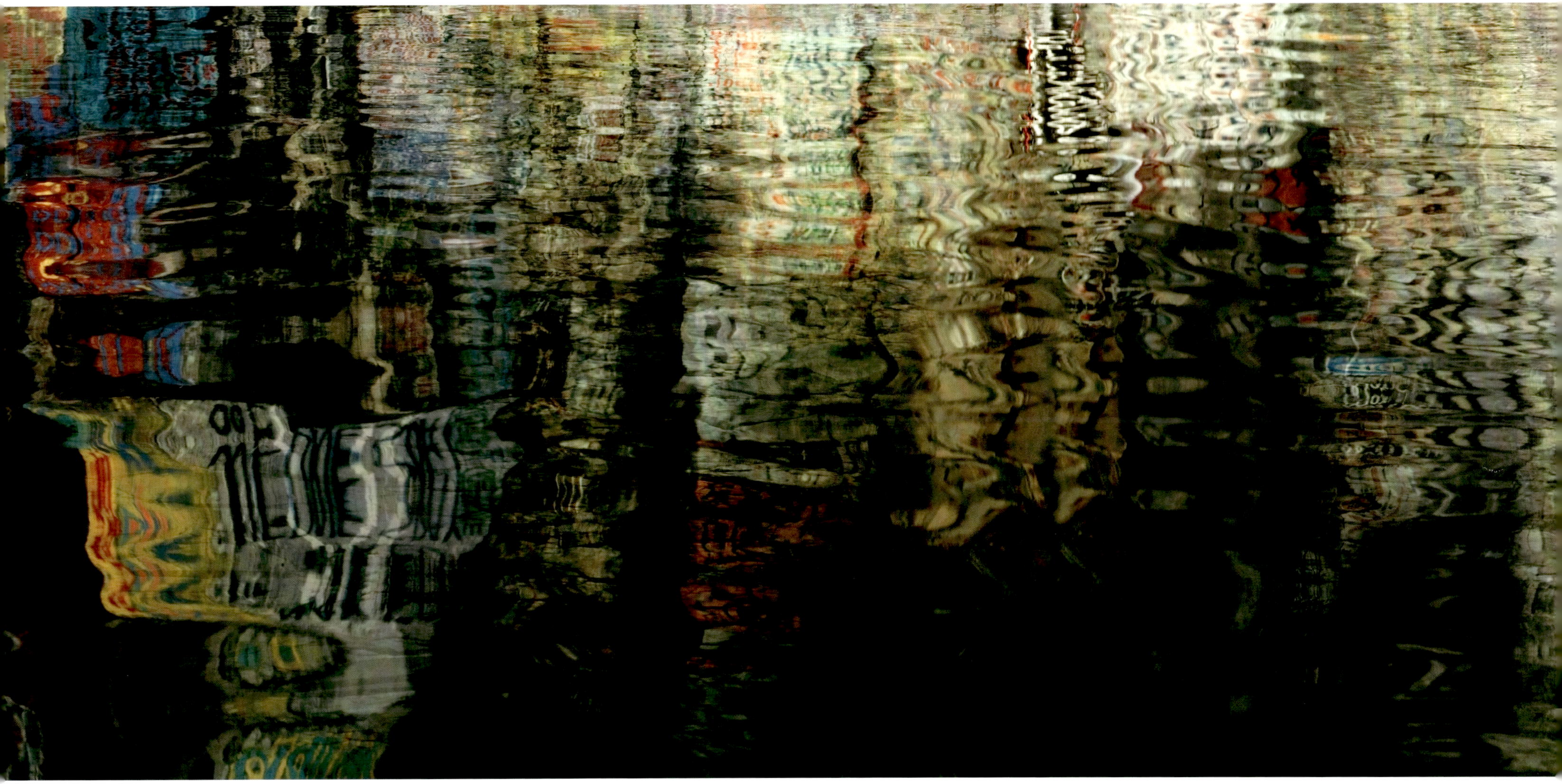

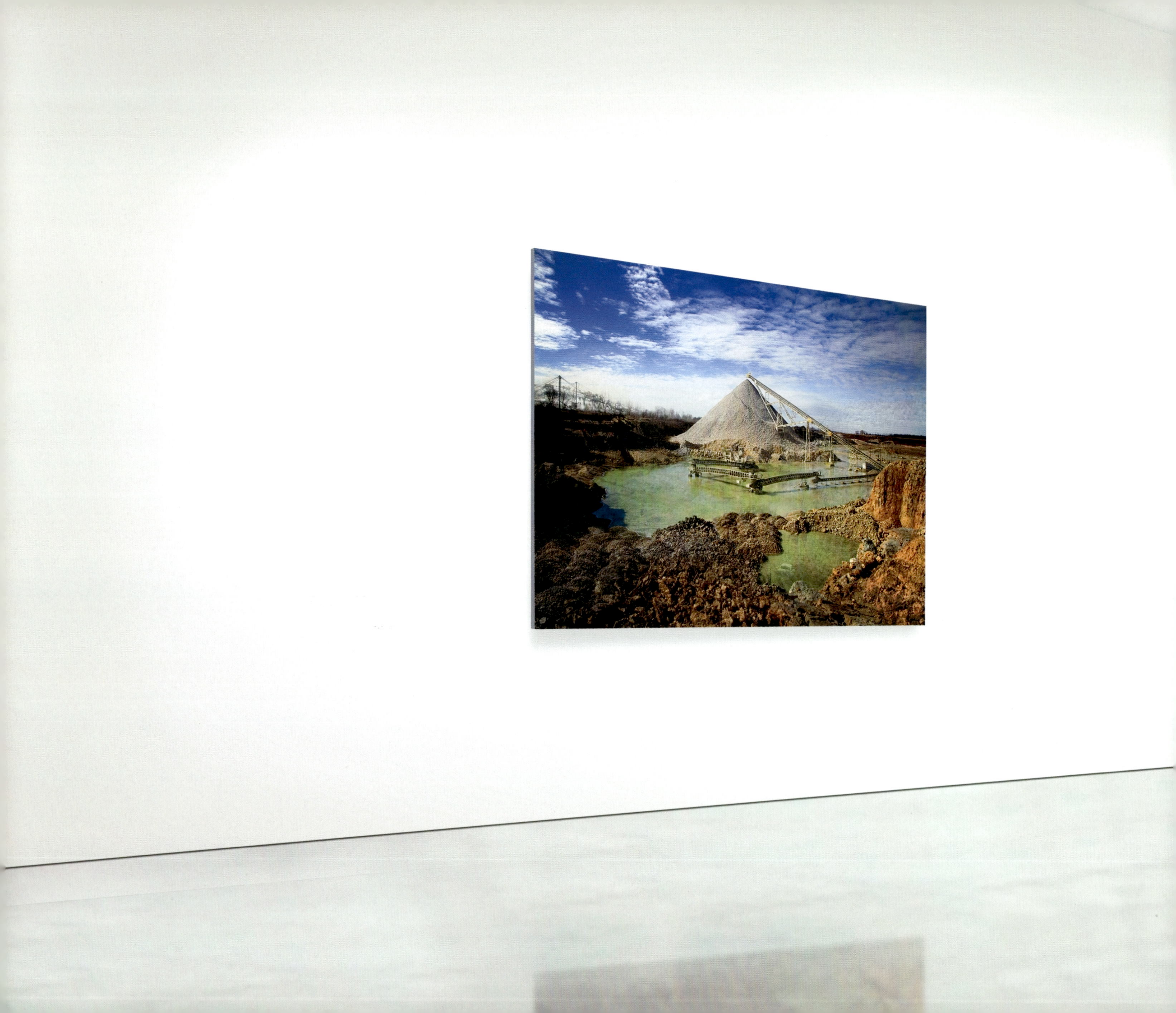

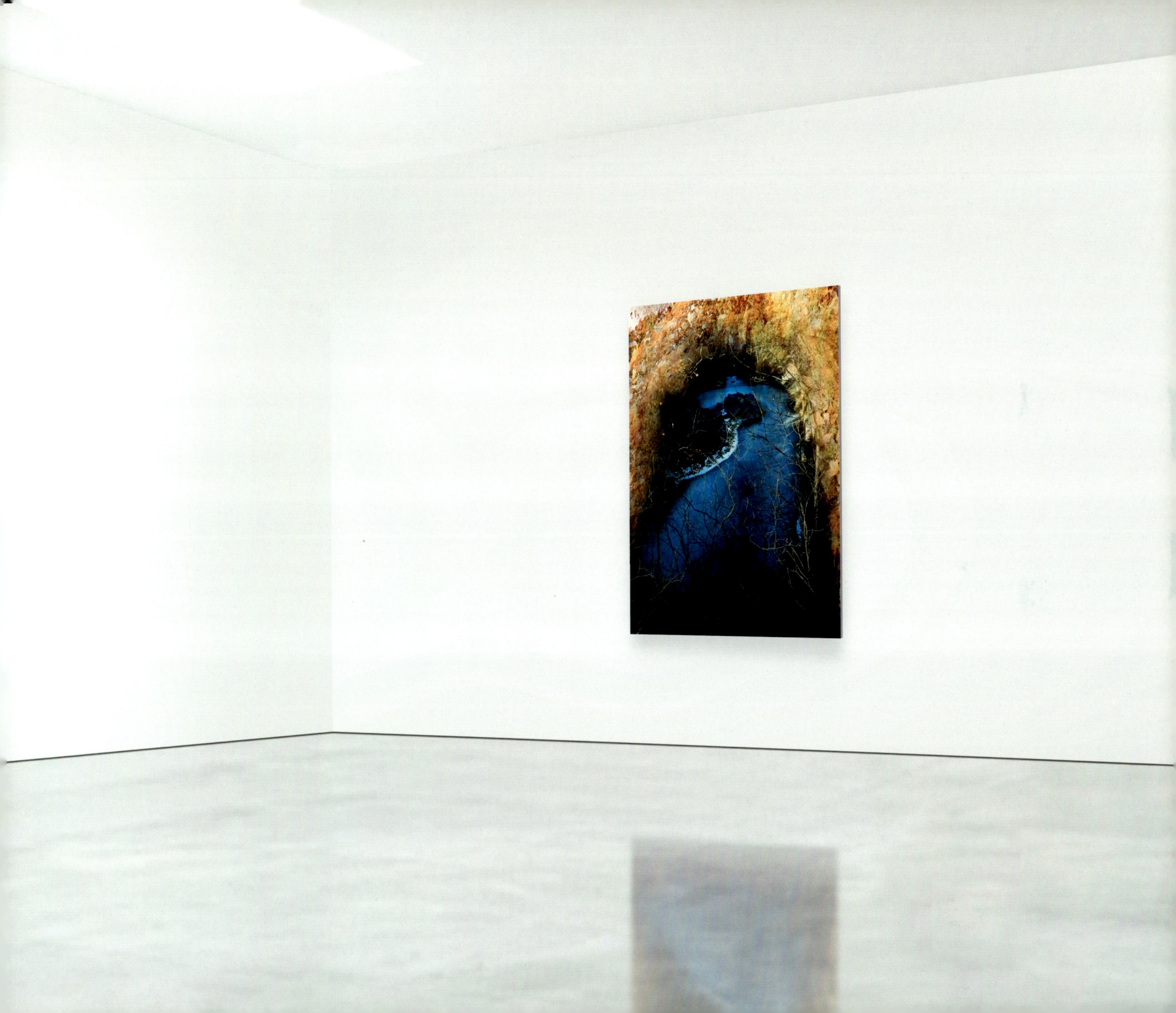

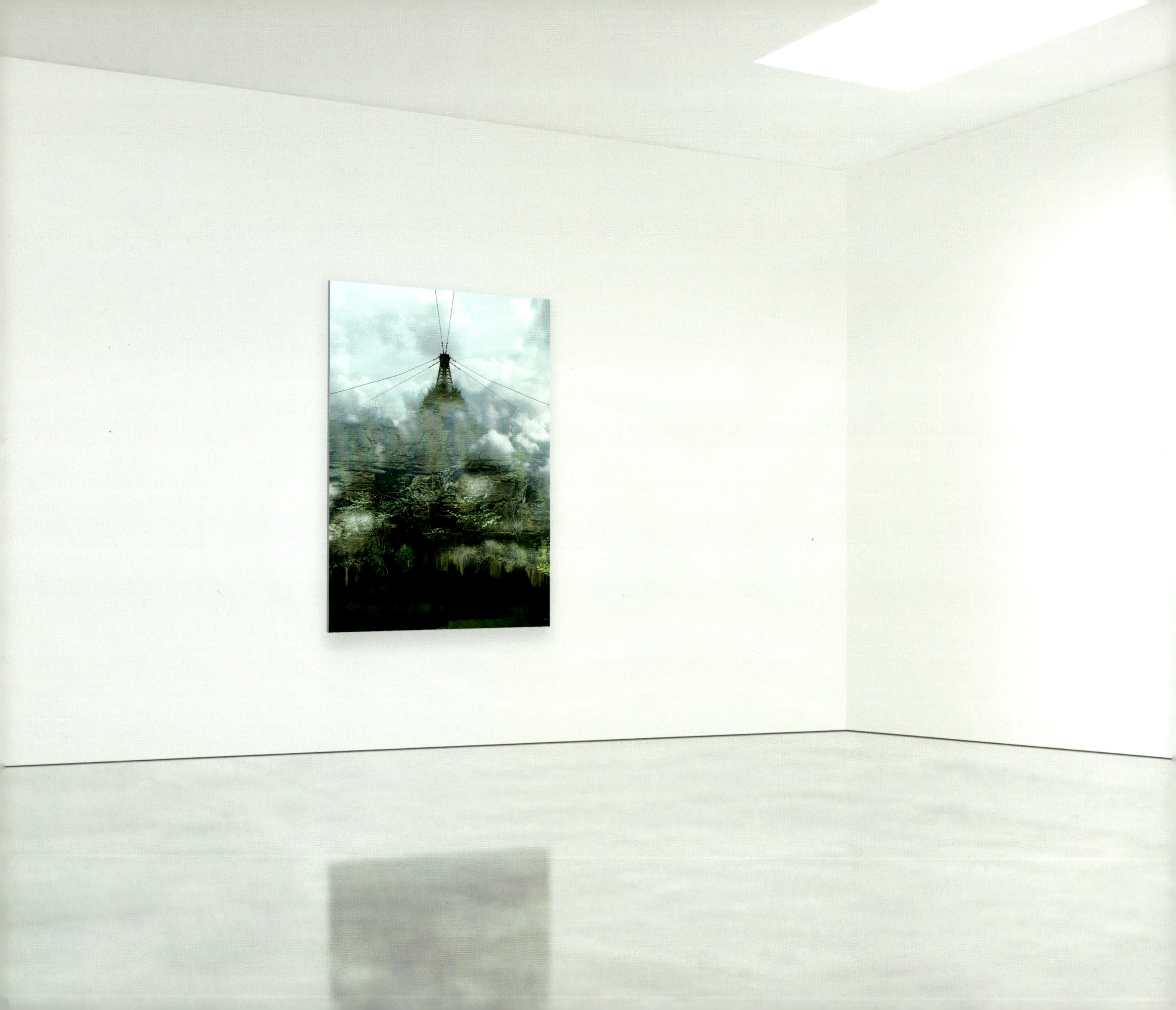

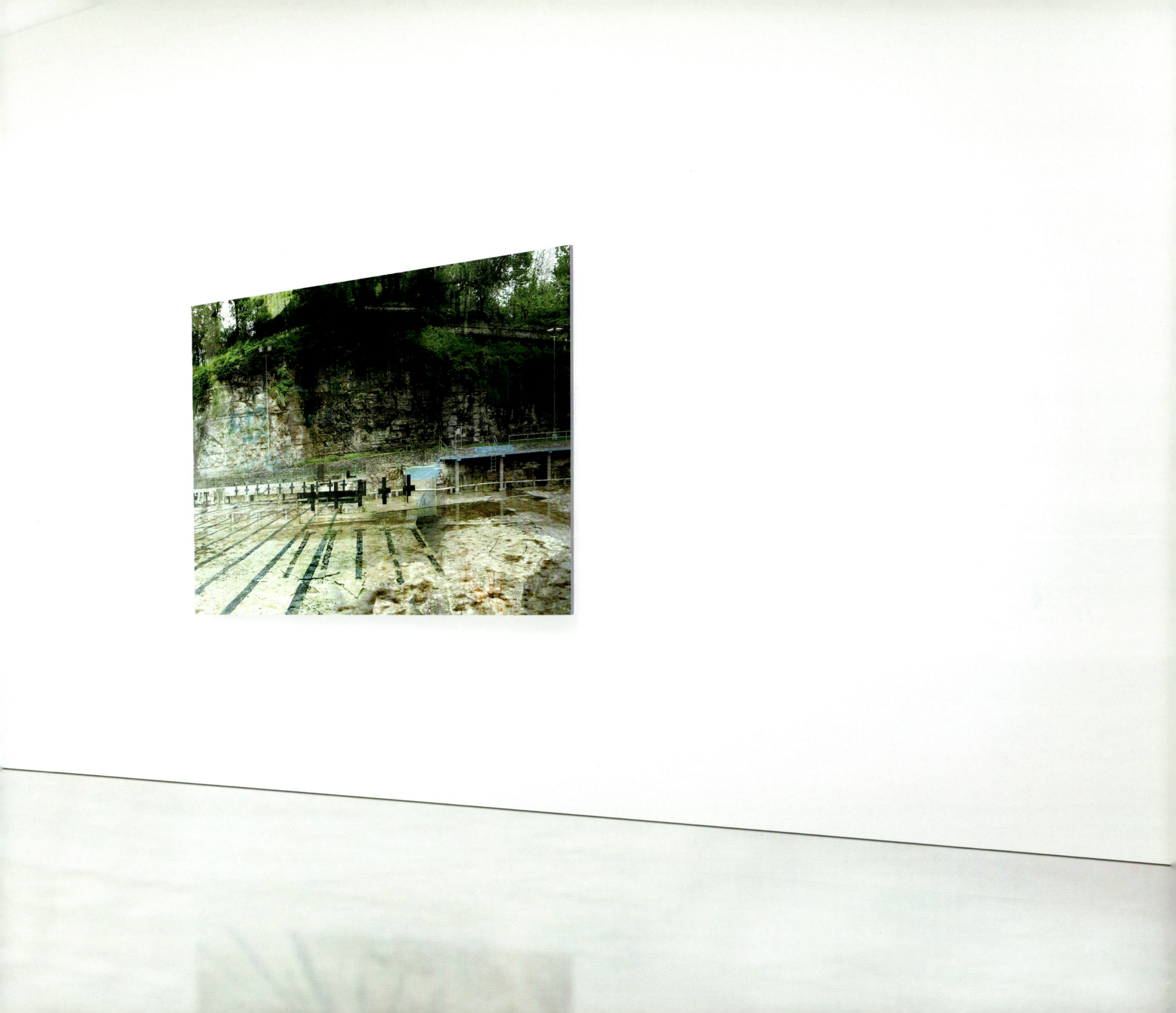

Elena Dorfman is a Los Angeles based photographer. Her previous works have explored the cultural, social, and sexual practices of marginalized communities. Published works from selected exhibitions include *The Pleasure Park*, photographic stills and 3-channel video, 2009; *Fandomania: Characters & Cosplay* (Aperture, 2007); and *Still Lovers* (Channel Photographics, 2005).

crump DAMIANI

in association with

MODERNISM

The first edition of Elena Dorfman: Empire Falling is published by

Damiani srl · Via Zanardi 376, Bologna 40131, Italy
t. +39 051 635 6811 / f. +39 051 634 7188
info@damianieditore.com
www.damianieditore.com

Book design by Traction, Cincinnati, OH

This book is part of a series of new titles edited by James Crump and is published in conjunction with the exhibition *Elena Dorfman: Empire Falling* at Modernism Gallery, San Francisco, January and February 2013.

ISBN 978-88-6208-266 2

Printed in January 2013 by Grafiche Damiani, Italy

Elena Dorfman would like to gratefully acknowledge Sarah Lyon, James Crump, Kevin Moore, Martin Muller and Danielle Beaulieu at Modernism, Andrea Albertini, Eleonora Pasqui, Alex Galan, Eoin Waxel, Ben Lord, JC Fernandez, Judith Turner-Yamamoto, Lila Lee, Wendy Earl, Bruce Halpryn and Chas Riebe, Heather Byer, and Paul O'Bryan.

Cover, Empire Falling 1, 2011, 50 x 33 inches

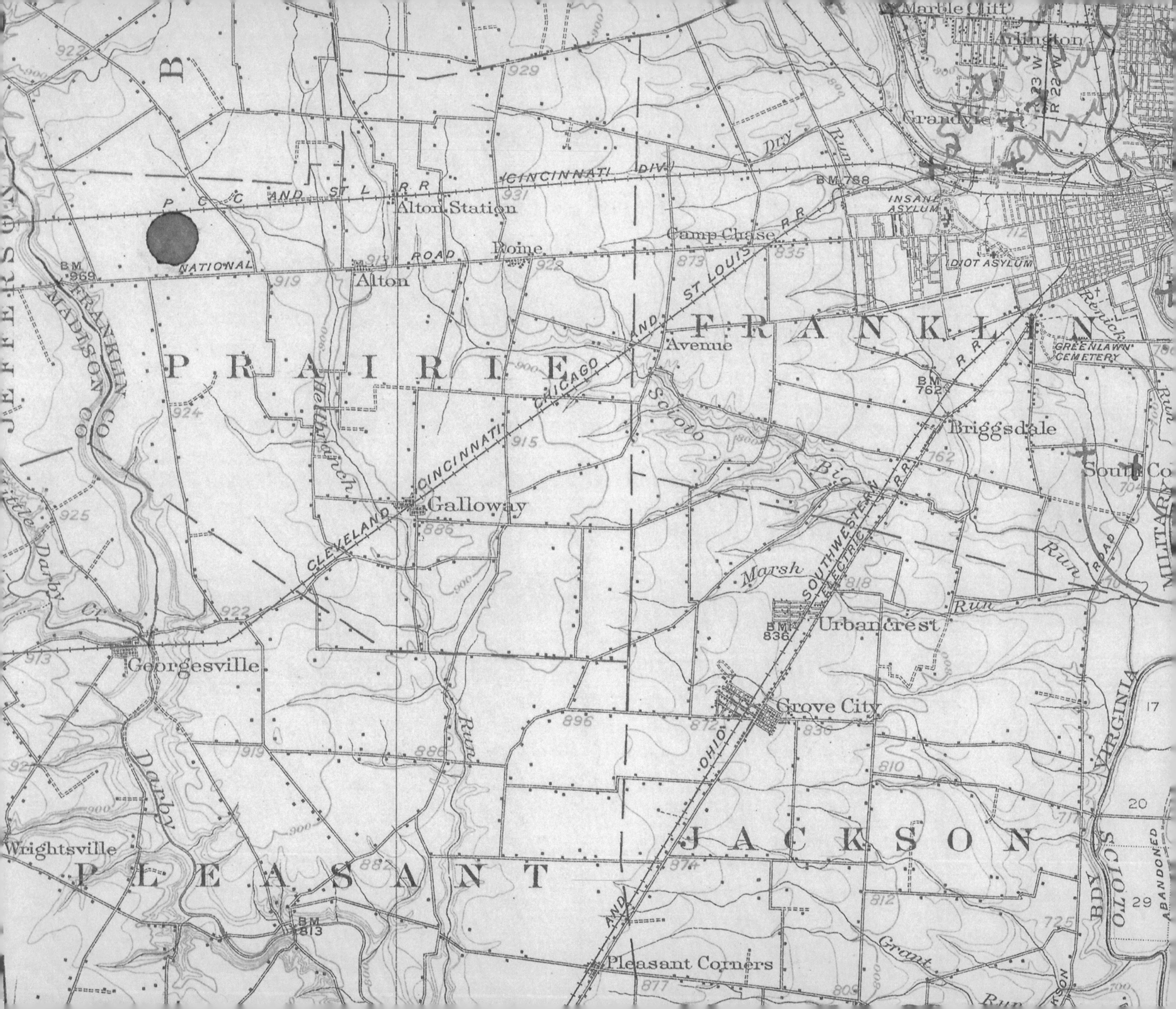

Marble Cliff
Arlington
Grandview
R 23 W
R 22 W
Dry Run
P C C AND ST L R R
CINCINNATI DIV
BM 788
Alton Station
INSANE ASYLUM
Camp Chase
Rome
NATIONAL ROAD
Alton
IDIOT ASYLUM
BM 969
JEFFERSON
MADISON CO
FRANKLIN CO
CHICAGO AND ST LOUIS R R
CLEVELAND CINCINNATI
FRANKLIN
Avenue
Renick
GREENLAWN CEMETERY
PRAIRIE
Hellbranch
Scioto
BM 762
Briggsdale
Big Run
South Co
Galloway
MILITARY ROAD
Little Darby Cr
SOUTHWESTERN ELECTRIC R R
Marsh Run
Georgesville
BM 836
Urbancrest
Grove City
OHIO AND
17
20
29
Darby
Wrightsville
JACKSON
PLEASANT
VIRGINIA
BDY SCIOTO
ABANDONED
BM 813
Grant Run
Pleasant Corners